Human Culture
A Moment in Evolution

HUMAN CULTURE
A Moment in Evolution

Theodosius. Dobzhansky
and Ernest Boesiger

Edited and completed by
Bruce Wallace

Illustrations by
Hans Erni

Columbia University Press/New York/1983

Library of Congress Cataloging in Publication Data

Dobzhansky, Theodosius Grigorievich, 1900–1975.
  Human culture.

  Bibliography: p.
  Includes index.
  1. Human evolution—Addresses, essays, lectures.
  2. Social evolution—Addresses, essays, lectures.
  I. Boesiger, Ernest. II. Wallace, Bruce, 1920–
  III. Title.
  GN281.D63 1983   573.2   82-22172
  ISBN 0-231-05632-X

Columbia University Press
New York Guildford, Surrey

Clothbound editions of Columbia University Press books are Smyth-sewn and
printed on permanent and durable acid-free paper.

Contents

Preface

On January 3, 1974, Professor Dobzhansky wrote Professor Etienne Wolff of the Collège de France in Paris, accepting an invitation to offer a course to be entitled "Problems of Organic and Human Evolution." He continued, "Although I am not able to lecture in French freely, I hope that my friend Professor Boesiger of Montpellier will have them translated and typed suitably for my oral delivery." The lectures alluded to in this letter were delivered in mid-November 1974.

The lectures that Dobzhansky delivered at the Collège de France were, through an understanding with his friend Ernest Boesiger, to form the basis of a joint, book-length manuscript. This is already clear on June 10, 1974, when Dobzhansky in a letter to Boesiger wrote of Grassé's *L'Évolution du Vivant*: "We must study this book, not to convince Grassé, which is obviously impossible, but to put in our future book arguments against his arguments." Their joint book, according to Dobzhansky, should "give the reader a basis for critical evaluation if he is familiar with Grassé's work."

On August 30, 1975, Ernest Boesiger died suddenly while attending the Fourteenth International Conference of Ethology. In a poignant note to Mrs. Boesiger, Dobzhansky notes how, as one grows older, one finds "more and more often that one's friends and acquaintances are no longer living." The truth of this statement was brought home to many population biologists several months later when, on December 18, 1975, Dobzhansky died. After Boesiger's death he had mentioned that "Ernest must have left some manuscripts for our projected book" and had asked of Mrs. Boesiger, "Will this book be lost?"

The book has not been lost. Mrs. Boesiger asked me if I would complete the unfinished manuscript left by her husband and Dobzhansky, and I agreed to do so. I have made every effort to complete the book as I felt they would have written it themselves: the book, after all, is by Dobzhansky and Boesiger. Dobzhansky was my teacher and friend for more than thirty years; Boesiger was an acquaintance, but one who arranged for the French edition of my own book, *Topics in Population Genetics*. To put the final touches (which, incidentally, included translating Boesiger's chapters from French into English!) on a manuscript prepared by old friends did not seem unduly onerous. As the lad on the old Christmas card says, "He's not heavy, Father; he's my brother."

Having said that this is their book—that is, Dobzhansky's and Boesiger's—let me admit at once that, left to my own devices, I would never have undertaken the task they accepted voluntarily. The reason is simply one of personality. Anyone who is familiar with my own work knows that I am most comfortable with research aimed at providing answers to rather precisely posed questions. When I speculate (as I have done with respect to the physical nature of gene control mechanisms), I do so most cheerfully when the speculations are obviously subject to experimental test by means of existing; or at least feasible, analytical techniques. Dobzhansky and Boesiger take a much more cosmic approach to problems. No picayune technical problem is to be discussed in this book; here we take on the majestic sweep of human evolution, from the lowly creatures who have as yet not been positively identified as our ancestors to remote descendants who, depending on the success or failure of cultural feedback mechanisms, may be dwelling in paradise or its opposite—or, not at all.

Two commonly used terms, counterintuitive and speculative, have meanings that are remarkably dependent on the mental bent of the one who uses them. Some years ago there appeared an article explaining the counterintuitive consequences of public housing for the poor: such housing increases the number of poor families in a community. The resident poor families having been moved into the newly constructed public housing, other—still poorer—families move into their now empty dwellings. Hordes of ecological biologists who are conversant with complex problems whose second-, third-, and even fourth-level ramifications cannot be safely ignored, would have foreseen this result of public housing immediately; students of ecology, with no prior warning, are expected to foresee such remote effects while taking both oral and written examinations.

Dobzhansky did not enjoy speculation, at least as he defined this term.

He did, however, immensely enjoy the interplay of ideas—pushing and pulling at the logic of any idea to see how far its consequences could be carried. Because intellectually he was head and shoulders above most of his colleagues, I believe Dobzhansky (and Boesiger, as well) enjoyed examining the implications of his research with respect to the nature and future of man. Perhaps I am being a bit unfair (after all, in his laboratory was a formally framed dime that he won from H. J. Muller by correctly predicting the outcome of a jointly designed experiment), but I suspect that Dobzhansky had much less love for speculations that could immediately be put to the test than he had for the grander variety, for pronouncements that suggested how the course of human evolution would—or should—progress.

One topic which I would have avoided like the plague is that which forms the basis of chapter 2: Natural Selection and Internal Teleology. Here I knowingly set myself in opposition to numerous first-rate biologists: Dobzhansky, Monod, Ayala, and Pittendrigh, to name a few. Still, I see nothing gained by saying that living things are unique in that they are purposeful. Or even by saying that tools created by man are created for a purpose.

My own view of the world is that problems exist which are in need of solutions and that furthermore, once it has been found, a solution need not be weighted down with another term such as "purpose." Four is the solution to the question, "What is two plus two?" But it is not the purpose of four to be that solution.

From the very earliest times, living (that is, successfully self-replicating) organisms faced problems that were inherent in the environment about them; some of these problems were passive ones arising simply from the nature of things, but others were active in the sense that they resided in or were caused by other living organisms.

Under such conditions, the successful sensing and interpretation by an organism of its surroundings could be of tremendous importance. Light and other electromagnetic waves as well as sound waves offer a means by which such an interpretation may be made. Thus, I do not feel compelled to say that the *purpose* of the eye is to see; rather, I see the eye and its associated interpretive region of the brain as a rather advanced device capable of reconstructing from reflected light waves the nature of their reflective sources, thus providing their owners with precious information regarding the environment.

Even in the case of man-made objects, I tend to avoid the word "purpose." Given the anatomical structure of human beings, how might one relieve one's feet by creating right angles at both knee and pelvic joints? Unless one is willing to roll back and gaze skyward, a support for the buttocks is obviously needed. Chairs provide one solution to this problem. Logs provide another. Low brick walls around college town buildings provide still another, as many university students will attest. On the other hand, unwanted sexual advances in the privacy of one's home may be discouraged by bashing a chair on the would-be rapist's head. Despite such use, no one claims that the purpose of a chair is to prevent rape. An assignment of purpose—and, hence, an extended discussion of teleology—is not a topic I would have entered into voluntarily.

Dobzhansky, in chapter 3 ("A Critique of Non-Darwinian Theories of Evolution"), spends a great deal of time discussing the "neutralist theory" of Kimura and others who hold that most observable molecular variants (in both protein and DNA molecules) are unimportant (that is, selectively neutral) to the survival and reproduction of their individual carriers. My own view differs somewhat from Dobzhansky's in this regard. In *Basic Population Genetics* (Wallace 1981) I argue that whether the neutralist theory proves eventually to be right or wrong, it is, *at this moment*, unnecessary. This argument is based on my belief that both of the reasons, the cost of natural selection and the size of genetic loads, cited by Kimura (1968) for stimulating the development of the neutralist theory are invalid. The arguments cannot be repeated here, but my claims are that the "cost of natural selection" is a computational artifact and that genetic loads are absorbed by deaths which are the inevitable fate of most zygotes of any generation of any population of any species. I must repeat, however, that my arguments do not reflect on the correctness of the neutralist hypothesis but only on the necessity for formulating it in the first place.

The fascinating matter of self-awareness arises in chapter 4, where the discussion includes this statement: "Scientific study of self-awareness is most difficult." I shall not differ violently with that assessment. Still, I feel compelled to record additional comments. Subsequent to Dobzhansky's death (at a time when it was believed that chimpanzees but not lower primates had an awareness of self), Professor B. F. Skinner and his colleagues have claimed an awareness of self among pigeons (Epstein et al. 1981).

I can muddy this aspect of the problem further by speaking of my cat, Red—a gentle, rather obese, orange-colored feline soul who comes equipped with extra toes. Over a number of years my daughter and a neighbor's son have trained Red to anticipate that, once they have confronted him with an arms-up, talon-handed threatening pose, this pose will be immediately abandoned if he snarls (a gentle snarl—sigh?—to be sure, but an audible one emanating from an open mouth). The training has been sufficiently effective that the open-mouthed response can be evoked by either child's making threatening gestures from across the street while Red, sitting on the back of the sofa, watches through the window. The children have now discovered that the snarling response can be evoked if the cat is in one room and the person exhibiting the threatening posture is in a second, out of sight except for a full-length closet-door mirror that allows eye contact between cat and child. Most surprising, they have learned that a person posing unseen behind the cat can evoke a snarl if the cat's shadow and that of the menacing person are on the wall in front of and visible to the cat; the cat snarls at the shadow on the wall. This account scarcely describes a well-designed experiment; however, if the cat can interpret a shadow as an immediate threat, might it not interpret its own shadow as "self"?

The discussion of self-awareness can be extended. Wherever it may have occurred during the earth's history, it surely represents one of the great advances in animal evolution; human beings, however, have taken a second, much more difficult, step: we are aware that others have self-awareness and that they, in turn, are aware of ours. Surely this second step must have preceded meaningful family conferences, tribal powwows, and intertribal deliberations. The concept of "do unto others . . ." of necessity, rests on an awareness that other individuals of one's sort possess their own self-awareness.

Among the items listed in nearly every inventory of traits needed for human evolution, along with proper physical size, an opposable thumb, stereoscopic vision, and a large brain, is the lack of estrus in human females. The continuous sexual receptivity of women to their mates is cited as one of the essential bases of permanent pair bonding, and of human family life. There are, one might point out, permanent pair bondings that do not depend on more or less continuous mating; geese, for example, form lifelong bonds with overt sexual activities limited to annual intervals.

Permanent bonding with a concomitant continuous sexual life makes sense, it seems to me, only if such activity is preceded, or at least accompanied, by an awareness of the feelings of others. What purpose would be served, for example, for male dogs and tom cats to remain permanently with their mates during the rearing of pups and kittens? The father's presence would serve no useful purpose—indeed, quite the reverse. Abandonment of estrus must have coevolved with the growing appreciation of another individual's feelings, with an awareness of another's self-awareness, with empathy.

To illustrate the tremendous gap between self-awareness and awareness of the self-awareness of others, I can cite the tribulations experienced by an American driving an automobile for the first time in Great Britain, where they drive on the "wrong" (left) side of the highway. For a few moments, the American driver may exhibit considerable confusion, especially when turning at intersections or negotiating roundabouts. Within two hours at most, however, the peculiarities associated with driving on the left are mastered: look to the right when entering traffic; keep to the left of the highway dividers. This quickly acquired confidence corresponds to self-awareness; in fact, it represents an awareness that one has successfully mastered the local tricks.

The next stage—the one corresponding to the appreciation that others have self-awareness—is exemplified by the distrust the now self-confident American exhibits toward other (British) drivers. For weeks or even months, he checks both right *and* left when entering traffic. This is not a casual check but a serious one based on a skepticism that others actually obey what to the American are peculiar regulations. Distant oncoming cars and lorries appear to be in his lane; they sort themselves out properly only during the last moments as they approach. The time and effort required to overcome a distrust of the driving patterns of native Britons illustrates the enormousness of the problem associated with behaving as if others appreciate one's own feelings, as if others are aware of one's own self-awareness. Ages of emotional travail must have transpired before an upraised empty hand became an accepted peaceful greeting—a sign of peace-be-with-you; those who first tried it may have quickly perished as a consequence.

This digression on self-awareness can be terminated by referring to still another sentence in chapter 4: "Self-awareness is the cause of another unique human quality: death awareness." The latter we must admit cannot exist without the former: that which will eventually die is me, my self. Still, much modern social criticism deals with the removal of death from the individual's immediate experience. Whereas self-awareness has been vulgarized into self-indulgence, self-gratification, and a deification of "Numero

Uno," death has been relegated to the narrow iron cots and small rooms of nursing homes and hospitals. Consequently, although the logical relationship of self- and death-awareness cannot be denied, the behavioral relationships of the two awarenesses are, I think, by no means as closely linked as Dobzhansky implies.

As one with a much narrower view of philsophical matters than Dobzhansky's, I shall express at least some misgivings over the discussion of ethics presented in chapter 4. The essence of the Darwinian view of living organisms is that organisms alone have certain characteristics and that these characteristics determine the nature of any population and, collectively, of any species. There is no *eidos*; there is no Platonic ideal which prescribes how organisms *should* be constructed: they are built, and they appear, as they are—period. I sense that this anti-Platonic view of living organisms did not displace in Dobzhansky's mind what seems to be his Platonic view of ethics. Dobzhansky raises the possibility in chapter 4, for example, that man is born either good or evil. At times, in discussing ethics and evolution, he shifts his comments from what *has been* or what *is* to what (in his opinion) *should be.*

Now, to discuss what should or should not be acceptable behaviors as if such decisions were made in outer space (where Plato's ideals reside, presumably) strikes me as risky business. Perhaps erroneously, I gain the impression from Dobzhansky's account that one's behavior is supposed to be governed, not by forces (or factors) arising from and responsive to the challenges of social interactions, but by a set of rules that have been engraved in stone. In this respect, I am probably much more sympathetic than Dobzhansky to Monod's claim that man, because he is beholden to no one, must eventually construct a code of ethics uninfluenced by myths. Despite the deep sympathy I have for Monod's view, I still fear the adoption of his recommendation. Why? Consider the havoc urban planners have wreaked on every major city in the United States. Look at the physical layout of the classroom buildings, sidewalks, and parking lots of any university campus, and then look at the spider web of paths that show where students *really* walk. The implementation of Monod's suggested ethic would take place so slowly, and in the process would involve the incorporation of so many erroneous, specious, and illogical reasons for this or that behavior, that in the end it would itself be a myth—or a religion: Monodtheism. Merely because a smoothly operating code of rational ethics may be difficult or nearly impossible to evolve does not, in my view, justify a belief in the existence of Platonic ideals of good or evil.

Apparently Dobzhansky, who authored chapter 7, was unaware that the studies of Sir Cyril Burt, which form a keystone for the types of data he cites in discussing the IQ's of twins reared together or apart, have come under deep suspicion (Hearnshaw 1979; see especially ch. 12). It appears that not only Burt's "experimental" data but also his loyal "collaborators" were entirely imaginary. While it is true that other studies on the genetics of IQ may have been carried out diligently and in good faith, we must recall a statement made by Sturtevant (1965:98) upon noting the complete turnabout in the nature of published data on the genetics of the ABO blood group system following Bernstein's (1925) demonstration that this system must be based on a three-allele (I^A, I^B, and i) system: "This tabulation [the contrast in observed frequencies of O, A, B, and AB blood types among children of O × AB marriages prior to and following Bernstein's three-allele model; the early data do not fit the model, whereas the later data do] raises some disturbing questions. One has the uncomfortable feeling that observers see and report only what they expect to find." Given Burt's status in the field of educational psychology, it is difficult, if not impossible, to assess the impact his fabricated publications have had on the interpretation, reporting, and publication of subsequent studies on IQ.

In discussing the high correlation between the IQs of identical twins reared together, Dobzhansky utilizes an argument which, although perhaps logically correct, strikes me as misleading or worse. Some persons, he notes, argue that the exceptionally high correlation between IQs of identical twins raised together may be caused by the similarity of their environments: Parents, teachers, and others treat identical twins more similarly than they treat fraternal ones. "This objection is invalid," he adds. If they are treated more similarly, it is because they *are* more similar: "the genetic constitution of a person is in part responsible for how that person is treated by others." The following example, utilizing the same reasoning, reveals its inherent weakness: Skin color has a genetic basis; skin color is the basis for racial discrimination against blacks; this discrimination has resulted in poor educational and employment opportunities; therefore, these inferior opportunities—and their measurable consequences, such as IQ scores—have a genetic basis. Following Dobzhansky's logic, the genetic basis for low IQ scores is all too clear: genes for black skin pigmentation are "in part responsible for how that person is treated by others." Nothing is gained by such reasoning; it merely misdirects our attention. The environment—in this case, the behavior of others—must under these circumstances be the only focus of attention. Too many persons, with punitive eugenic proposals in hand, await the assignment of "heritable" or "genetic" to any phenotype they consider undesirable; for geneticists to oblige these persons by arguing that unnecessary personal reactions to certain genotypes are in reality caused by those genotypes is unseemly.

I shall now reverse my overly critical stance and, to the possible amazement of early reviewers who tended to be critical of his speculations, defend Ernest Boesiger's account of the evolutionary origins of esthetics and a sense of beauty. Although questions remain about the linkages among prehominid fossils, no doubt remains in the minds of most biologists that modern humans evolved from lower forms. In so doing, it is inconceivable that at one moment a sense of beauty fell, like an overnight snow, and blanketed the mind of man. On the contrary, as Boesiger suggests, the sense of beauty—esthetics—has developed gradually over a long period of time. A sense of beauty coevolved with the mental capacities of man's ancestors. In many respects, the evolution of a sense of beauty can be thought of in terms similar to those mentioned with respect to the mutual feelings of dependency between husband and wife. I said earlier that permanent pair bonding depends on one's awareness of the self-awareness of others. At this point, Boesiger might well feel compelled to point to the enormous variation exhibited by different human beings: some have no sense of beauty; some have (or are) battered wives and abused children.

My intent in writing this Preface has not been to criticize the manuscript that I have completed; reviewers will undoubtedly relish that task, while vigorously criticizing my small contributions (the Introduction and Chapters 1 and 8) as well. To add my name to those of Dobzhansky and Boesiger in writing against those who would once again impose on the educational systems of the various states of the United States a religious doctrine of man's origins in the guise of biological science has been a privilege. Even as these words are written, a president of the United States, Ronald Reagan, claims that having "studied" the matter of evolution, he finds many flaws in the theory and, therefore, favors the inclusion of creationist theories in the public school science curriculum. Presidential politics being what they are in the United States, deciding whether or not to take Mr. Reagan's words se-

riously is difficult. Is he simply appealing to an already committed fundamentalist constituency? Or does he believe that every gap in man's scientific
knowledge must retain a standby, "God's will," as a possible explanation? In
reading of the deaths of war veterans following the American Legion's 1976
convention in Philadelphia, a few religious fanatics did, in fact, suggest that
these deaths were caused by a disease newly created by God. Many months
elapsed before the true nature of the disease was determined. Rest assured,
the gaps that remain in the theory of evolution will also be closed. Creationists, like their colleagues who believe in a flat earth, have no available
option except to retreat, fabricate, misquote, deny, and retreat once more.

It is a pleasure to acknowledge here the great help provided to me by Dr.
June Nasrallah, who double-checked my translations of Ernest Boesiger's
two chapters and, in doing so, discovered inadvertent errors—some of which
might better have been retained for levity's sake. Thank you, June. And
Cathy Tompkins—thank you for preparing the typescript. Both Mrs.
Boesiger and I wish to express our gratitude to Ernest's friend Hans Erni,
who created the illustrations for this book.

Blacksburg, Va.   Bruce Wallace
February 15, 1982

Introduction

This is a book about evolution, especially the evolution of man. It is, however, not a *Handbuch* in the German sense; those seeking descriptions of the latest paleontological and archeological discoveries should look elsewhere. Here, we shall treat of the logic of evolution, and of the status of the human species (*Homo sapiens*) with respect to evolutionary change. Human beings have not always been as they are; nor will they continue to be as they are in the future. Ancient animals that were ancestral to our species are at some point denied our species' label; the same shall be true of our remote descendants. Stability and change are matters of perception. We can speak of *Homo sapiens* yesterday, today, and tomorrow only in the same sense that we can speak of a point of land known as Land's End, England, or another known as Dubrovnik, Yugoslavia. At some future time neither of these named localities will exist. Those who are familiar with the sunken cities of ancient Greece, or with the Mediterranean ports of that same era which now lie buried in desert sands several kilometers inland from Alexandria, Egypt, or with the virtually annual crumbling of the Adriatic coastline of Yugoslavia, appreciate the speed with which changes can occur in the earth's physical appearance. Racial differences among large segments of mankind bring a similar appreciation to the student of human evolution. Still, for many, if not most, persons the nature of mankind is fixed and racial differences represent a biological constant. To these persons it is as inconceivable that the human race will change (or has changed in the past) as it is that Honolulu should be anywhere but 2,400 miles southwest of San Francisco.

The biological sciences have undergone a tremendous revolution during the past two or three decades. Molecular biologists armed with technical procedures and tools unheard of only a short time ago are determining the precise chemical composition of even the most complex of the macromolecules required for life. The purine and pyrimidine sequences of DNA (deoxyribonucleic acid) and RNA (ribonucleic acid) as well as the sequences of amino acids in proteins are now revealed by automatic or semiautomatic scientific apparatus. The educational backgrounds of the younger of the modern research scientists are quite different from those of older biologists. To a large degree, the two generations (or two schools, since actual ages may not be all that different) speak different languages. Elaborate and at times tenuous arguments learned by members of the older school have been bypassed and ignored by those of the younger one. Not that this constitutes a previously unheard-of circumstance in biology. No one spoke of Mendel's work on garden peas for nearly thirty-five years; his paper now marks the origin of the science of genetics. Calvin Bridges, a pioneer *Drosophila* geneticist, wrote at least one research paper in which the name of his research organism was not even mentioned; that was an act of supreme confidence. It was also an act that separated the outsiders from the insiders in the field of genetics.

The periodic changing of the guard that accompanies successive technical revolutions in biological research leads to cyclic views concerning the large, unsolved problems of that science. The chemists and physicists who teach research techniques to the young biologists of one generation had their own biological training in an earlier generation. Thus, it seems that every twenty-five or thirty years once-discarded ideas are resurrected, refurbished, and recast as the avant-garde notions of the day. So it is that Richard Goldschmidt's hopeful monster—an abnormal, mutation-bearing individual in search of a similar mate—is cited in the 1970s as a possible factor in biological evolution, just as Goldschmidt suggested during the 1930s. Similarly, many persons have turned from natural selection as a molding force in evolution; instead, they seek (as chemists are wont to do) nonhistorical explanations for what is the most obvious of all historical processes—gradual change among living things with the passing of generations. Grassé, the modern-day French encyclopedist, even while asserting that only paleontologists can speak of evolution, turns to newly assembled DNA sequences for a scientific "explanation"; in making his opening assertion, Grassé forgets that Cuvier, a paleontologist, was an ardent antievolutionist. The facts of paleontology (in company with those of many other biological sciences) do not "prove" evolution—especially to persons who refuse to believe.

The study of evolution, as is made clear in a later chapter, is no longer a matter solely of pedantic interest. The world *is* changing; even the most confirmed statists can sense that. Modern technology appears to thrive on agents that are noxious to life. This applies not only to the X rays of modern industry and medicine but also to the radioactive contaminants generated by nuclear power plants and to the genetically harmful compounds produced in carload lots by the chemical industry. Productive farmland, essential for the proper nutrition of the world population, disappears beneath asphalt in the construction of homes and factories. This is true not only in Urbana, Illinois, and Lancaster, Pennsylvania, but in the Nile delta of Egypt as well. In the latter case, a nation of 40 million persons possesses only several hundred square miles of arable land; nevertheless this land is sacrificed to construction projects under the guise of progress.

Who will speak out in defense of those resources that are necessary for life? Who will fight against recklessly mortgaging the future of the human species and that of the other species on which man's existence depends? Not those who believe our stay on earth is a matter of God's will. Not those who wring their hands while denying a precise understanding of how evolution (including extinction) works. No, the fight for the future of man will fall primarily to those who understand man's origins, man's frailties, the dynamics of the birth and death processes, the meaning of and requirements for population stability, and the constraints under which evolutionary change can occur.

The fundamentalist minister tells his TV audience that he doesn't care if hamburger is $3.95 per ounce. And that he doesn't care if there should be another war. In truth, he doesn't care about people—only their souls. And so, the facts of evolution and of evolutionary processes must be taught in order to train the caretakers of mankind, to build a society in which every person is brother or sister to every other person, and to create an ethic in which provision is made for the generations of brothers and sisters yet to come.

Human Culture
A Moment in Evolution

CHAPTER 1
Different Forms of Natural Selection

A population of living things, plant or animal, tends to take on the characteristics of its reproducing members. This is true, for instance, of a population of bacteria. If a population of these bacteria encounter an antibiotic such as penicillin, nearly all will die. A few individuals carrying a rare mutant gene may survive, however. Within hours, the original population of bacteria will be restored (at twenty minutes per generation, only twelve hours are needed for the progeny of a single mutant individual to overwhelm any normal host) to its original numbers. These individuals will treat the presence or absence of the formerly fatal antibiotic with amazing indifference. Let us repeat the opening sentence: *a population of living things tends to take on the characteristics of its reproducing members*. No matter how many Smiths live in a town, if only the Browns bear children, Smiths will eventually be omitted from the city directory while the Browns take over.

The consequences of differential reproduction in a sexually reproducing population are obvious. Nevertheless, many deny such consequences. Why? First, because they regard all individuals as being genetically identical. Today this statement should be revised: these persons may regard all individuals as being genetically equivalent but not necessarily genetically identical. Following a long period during which geneticists thought that there was one, and only one, *normal* gene at each locus for a given species, many geneticists now believe that there are a series of functionally equivalent (selectively neutral) alleles at each locus. Genetic diversity, in other words, need not require diverse positive selective forces—only selective indifference and reproductive accidents.

Ignoring selectively indifferent or neutral genetic variation for the moment, the following three conditions are both necessary and sufficient for the occurrence of evolutionary changes within populations.

1) The members of populations (even sibs from the same progeny) must differ from one another with respect to a trait (or traits) that is important for survival and ultimate reproduction.

2) The differences between individuals both within and between sibships with respect to this trait must be, at least in part, genetic: like, in this instance, must beget like.

3) In the process of reproducing, the population of one generation must produce more offspring than can successfully survive to become parents of the next generation.

Given the three conditions listed above, evolutionary change—that is, a systematic change in the genetic composition of a population with the passage of successive generations—is not just possible: it is inevitable.

Evolution involves a change in the genetic composition of populations. Even in the case of adaptively important traits that are controlled by single gene loci, the interaction of natural selection and gene action can be rather subtle. For example, of the nearly 100 species of moths that have successfully adapted to the heavily soot-polluted areas of Britain and Germany by becoming darker, all but one or two have become so as the result of dominant or partially dominant alleles. Why this bias toward dominant alleles, rather than recessive ones? Because every initially rare dark individual that survived to reproduce because its dark color had a dominant genetic basis transmitted the responsible gene to one half of its progeny, thus offering them an identical protection. A rare dark individual that owed its color to the possession of two recessive alleles might also escape predation only to produce light-colored offspring by mating with one of the more common light-colored forms. Evolution proceeds most rapidly when offspring resemble their selectively favored parents. On the other hand, the mechanisms of inheritance among higher forms of life (higher plants, flies, mice, and human beings) provide considerable inertia to the demands of selection because of the segregation and recombination of hidden recessive alleles: higher organisms evolve slowly, while retaining the ability to reverse their direction of change should the responsible aspects of the environment themselves undergo a reversal.

Directional and Diversifying Selection

Because of our dependence on domesticated agricultural plants and animals for the bulk of our daily food, most persons think of selection in terms of artificial selection and of artificial selection in terms of directional selection: more milk per cow, more grain per acre, more eggs per hen, more bacon per hog, and more of a host of other items.

Certainly the artificial selection that has followed the domestication of plants and animals, some twelve thousand years ago, has been largely directional. In more recent years, the experimental work that has been carried out both with food crops and animals and with more convenient laboratory organisms has dealt primarily with the question, "How can more be obtained with less effort?"

To believe, however, that selection overall has been entirely unidirectional would require that the diversifying effects of man's husbandry practices be entirely overlooked. Consider, for example, the numerous breeds of dogs; shepherds, retrievers, bull terriers, toys, and others. Through the isolation of human settlements, or through the isolation of breeding stocks, the species *Canis familiaris* has been rent asunder by a series of directional selections each of which proceeded on its own path—some leading to animals capable of fighting marauding wolves, others to animals capable of attacking badgers in their burrows. The same sort of pattern can be seen in horses, where a series of draft animals (Clydesdales, Percherons, and Belgians) have been developed for farm work and another series (quarter horses and Arabian steeds) for riding. Cattle are notorious among animal breeders—breeds have been developed for any one, two, or all of three useful traits: milk, beef, and transportation.

Many modern agricultural specialists are gravely concerned over the loss of genetic diversity that resides for the moment among the many breeds and varieties of animals and plants that have been developed through centuries and millennia of diversifying artificial selection. This loss is especially severe in the case of cattle, where local breeds are rapidly displaced by the economically more efficient breeds and where the retention of even a small herd of less efficient animals imposes an intolerable burden on the private dairy farmer. The solution to this loss of germ plasm may well be the apportionment of the vanishing breeds among the many well-run zoological gardens of the world, where each zoo (perhaps for the amusement and education of children) would receive a government subsidy allowing it to perpetuate one or more now exotic, and endangered, breeds of cattle.

A species, such as horses, becomes fragmented into disparate breeds through diversifying selection; within each breed, however, the overall aim of the breeders has been directional. This is not to say that Clydesdale and Belgian draft horses have not undergone diversifying selection for the prevailing climates of two dissimilar geographic localities, or for resistance to local intestinal worms and other parasites; selection has been directional with respect to the size and conformation of the body which enables the animal to perform heavy labor. These are draft animals.

Any wild species of plants and animals is similarly exposed to selective forces that pull local populations in diverse directions: many localities are crazy quilts of dry and moist patches of ground; the quality of the soil with respect to texture and nutrients is also quite often patchy; other species of plants and animals (competitors and grazers) themselves have patchy distributions. The outcome of such natural conflicts and selective inconsistencies can be the restriction of the species to one type of patch only, the establishment of a polymorphism maintained by segregating alleles, the development of a plastic phenotype that permits each individual to respond properly to its immediate environment, or (in a sense, a repeat of the first outcome mentioned) the splitting of the one species into two, each one specialized to cope with one of the common types of environment. The most spectacular outcome arising from the formation of specialized species can be seen in the massive adaptive radiations of various animal groups, such as the giant lizards of past geologic ages or the marsupial mammals of Australia and New Zealand.

Normalizing Selection

Diversifying and directional selection can be discussed together because, once the divergence of selected populations has gotten under way, the chief tendency is along a single direction within any single line. The truth is somewhat more complex, of course, because any population—even a domesticated one—faces a multitude of problems, to which individuals respond differentially, besides the one problem most obvious to the observer. Indeed, an educated liver fluke might overlook the fine wool of Merino sheep while marveling at these animals' resistance to various parasitic hazards encountered in Spain.

Normalizing selection is quite another matter: it is selection that tends to favor the average or near average individual of the population. Such selection places a premium on the status quo. Most natural selection is of this sort. If, for example, directional selection had added even one millimeter per generation to the average height of human beings over the past ten thousand years (four hundred generations), we should now be 0.4 meters, or some 16 inches, taller than our rather remote ancestors. Modern man is not that much taller, on the average, than was Cro-Magnon man; despite demonstrable genetic variation in the height of individual members of any human population, there seems to have been no directional selection with respect to this aspect of human morphology. Because extremely tall and extremely short persons tend to be regarded as "different" by their fellows, the average number of children produced by persons with these extreme phenotypes tends to be smaller than average: not only is selection not directional, it operates to the disadvantage of both extremes.

Natural selection works through the differential reproduction or survival (or both) of genetically dissimilar individuals. To demonstrate normalizing (or stabilizing) selection with respect to certain phenotypes is not difficult. The death rate of newborn babies is lowest for those weighing about 7½ pounds and higher for those weighing both less and more. The hatching success of duck eggs is greatest for middle-sized ones, while lighter and heavier eggs fail to hatch more often; when poultry breeders apply artificial selection to chickens in an effort to get larger eggs, they find that hens laying smaller-than-average eggs produce the most chicks. Fruit flies possessing intermediate numbers of small bristles on any one of their many body segments tend to leave more offspring than do those possessing bristles in smaller or greater numbers. Each of the traits mentioned in this brief list is known to be affected by the individual's genotype. A good deal of natural selection, consequently, tends to stabilize the population, to keep matters as they now are. "Don't rock the boat" could be a philosophy attributed to selection in natural populations.

The Interplay of Culture and Biology

Cultural evolution and biological evolution are frequently treated as if they were two independent events; sometimes they are treated as if they act in parallel, but more often they are treated as tandem events. The latter view pretends, for example, that man's biological evolution has come to an end, having terminated with the origin of human culture. To the extent that adjustments to changing seasons are now made largely by adjusting thermostats or by donning or discarding clothing, this view may be correct. Biological evolution has not ceased, however; furthermore, it may easily occur in response to a cultural event, and vice versa.

Human beings possess a variety of hemoglobin molecules; these are made under the direction of genes at different gene loci (the alpha-, beta-, gamma-, delta-, and epsilon-polypeptide chains); large numbers of variant alleles are known at each locus. Probably the most notorious (or best known, depending on one's point of vew) of the beta-chain alleles is the S (for sickling) allele. The beta chain of hemoglobin S contains a substitution of one amino acid (valine) for another (glutamic acid) in a chain of 146 amino acids. The consequences of this substitution are several: the production of hemoglobin molecules that form long needle-like crystals when deprived of oxygen (these crystals cause a characteristic deformation—a sickling—of red blood cells); the production of hemoglobin molecules that migrate in an electrical field at a rate that differs from that of normal hemoglobin, because the amino acid substitution alters the electrical charge of the molecule; and the production of a hemoglobin which, perhaps because of its tendency to distort the shape of the red blood cell, confers a resistance to malaria on persons who carry one (not two) S-alleles. Carriers of two S-alleles (sickle cell homozygotes) suffer a severe anemia which in former times, under the primitive conditions of equatorial Africa, was usually lethal.

To illustrate the operation of natural selection, consider a hypothetical community in a region of Africa where malaria is endemic. We postulate that at the outset only persons with normal (non-S) hemoglobin inhabit this community. One fourth of all newborn babies die of malaria before reaching maturity (a not unreasonable figure; all other sources of death are ignored in this discussion). In order for the community to continue its existence generation after generation without dwindling away because of malarial deaths, each mother must bear an average of 2⅔ children (this number would be considerably higher if we were to include deaths by dehydration, bacterial infections, weakening by parasitic infections, and other debilitating afflictions) three quarters of the 2⅔ children survive ($\frac{3}{4} \times \frac{8}{3} = 2$) to replace their mother and father in the population.

Now, let the S-allele enter this population either by mutation or by the arrival of S-bearing heterozygous carriers (persons carrying a single sickling gene each). The mutant allele does not affect its heterozygous carrier in any obvious way. (Such persons are identified by placing a drop of their blood on a sealed slide overnight and examining it under a high-power microscope the next day; neither slides, microscopes, nor a reason to examine blood samples exists in our hypothetical community.) Women who carry the S-allele (or women with normal hemoglobin who marry men carrying the S-allele) will produce babies of whom one half will have normal hemoglobin and one half will be carriers of the S-allele. The latter are immune to malaria and survive; one quarter of the former die from malaria. Because each woman, on the average, produces 2⅔ babies (of which two usually survive), women heterozygous for the S-allele (or whose husbands are heterozygous for this allele) will produce 2⅓ surviving children (one normal child plus 1⅓ S-bearing ones).

With no knowledge on the part of the natives as to what is happening, the newly arisen S-allele will spread in the population: the heterozygous carriers survive bouts with malaria, and one half of their children also survive. Natural selection acting alone, with no cultural interference, will lead to the establishment of the S-allele *at high frequency* in the population. As the allele grows in frequency, the villagers will note increasing numbers of children afflicted with and dying of a previously unknown, painful, debilitating anemia that is accompanied by several characteristic skeletal abnormalities.

The process described above, although controlled entirely by natural forces, would have cultural ramifications. One, for example, would concern the care required by children afflicted with a severe disease never seen before in the entire history of the community. A second, which would become increasingly important with time, would involve the increase in the village population. At the outset we described a village population in which each mother left two surviving children, replacements for herself and her husband. Heterozygous carriers of the S-allele produced (when the allele was rare) 2⅓ surviving children—more than enough to replace the parents. Even when many (about 4 of every 100) babies are dying of the anemia caused by the double dose of S-alleles, the average number of surviving children per couple will be approximately 2.13. No population can cope with a 6-percent increase in size for long; excess individuals will migrate to other localities, or causes of death other than malaria will increase in frequency, or families will in some manner lower the average number of children they bear. Judging from events in today's overpopulated nations, the last-mentioned possibility would be the least likely to be adopted by the population.

A second example of natural selection, almost as purely biological as the first, can be built around the theme of one of Hardin's (1963) essays: "Blessed are the women who are irregular, for their daughters shall inherit the earth."

Hardin's essay describes the events that would follow if all forms of birth control other than the rhythm method were outlawed. The rhythm method is based on the supposition that the menstrual cycles of women are regular cycles—related, perhaps, to the phases of the moon, as the term implies. Women, however, are not all that regular in their monthly cycles; a substantial number are markedly irregular.

Now, if the rhythm method of birth control were the only available means of avoiding pregnancy, the women with the most nearly regular cycles would have the fewest children. Accidental pregnancies would beset those women whose fertile periods occurred at unpredicted and unexpected times.

A woman's menstrual cycle is a physical phenomenen that is controlled by the interactions of several hormones and hormone-sensitive tissues. That such a cycle should possess a duration that is related to the phases of the moon suggests already that the cycle is a biological phenomenon subject to modification by natural selection—that is, it is under genetic control and this control responds to the demands of natural selection. Consequently, if irregular women produce more daughters each generation than do regular women, we can rest assured that these daughters of irregular women will in turn prove to have irregular periods. Should the rhythm method continue to be the only method of birth control, the regularity of women's monthly cycles would become a historical fact, not a fact of life. Birth control would fail.

Events that are now befalling daughters born of mothers who took DES (diethylstilbestrol) during pregnancy illustrate that Hardin's prediction is not farfetched. These young women—"DES babies"—have shown a number of exotic disorders (such as vaginal cancers) which might easily be blamed on the effect of the hormone on the developing fetus. More recently, it has been found that women who were carried as DES babies are now experiencing a variety of difficulties during childbearing; these troubles have also been blamed on the prenatal exposure of these women to DES. No compelling need to blame DES exists in the latter case, however. One need only recall that the mothers of these DES babies were treated with the hormone because they were having difficulties during pregnancy. DES babies, almost by definition, are babies born to women who under normal circumstances would have been unable to bear children, or at least to bear them easily. It should be no surprise, then, that the daughters of these women prove to have difficulties during their own pregnancies. The phrase "like begets like" was not coined without cause; on the contrary, it represents the culmination of shrewd observations made by human beings on themselves and their neighbors over the course of many centuries.

In the first example, we considered the establishment of a mutant gene (for hemoglobin S) in a population solely as a consequence of biological phenomena; the presence of this gene then altered the community's culture because of the severe anemia that affects its homozygous carriers and because of the growth in numbers of village inhabitants resulting from the lowered number of malarial deaths. The situation described by Hardin with respect to the rhythm method illustrates the immediate biological response to what was described as a culturally determined practice: an attempt to control births by a church-approved rhythm method. Culture, in this case, would provide a challenge to which natural selection would, in turn, produce a biological response (menstrual irregularity).

Another example, also largely hypothetical, can illustrate a postponed biological reaction to preceding cultural changes. This example is, in essence, the converse of that involving the sickling gene. In that case, as biological events ran their course, they eventually gave rise to matters requiring a cultural response (to illness and to overpopulation). Now we shall consider a case in which cultural events run their course, eventually giving rise to matters requiring biological responses.

During World War II, more than sixteen million persons served in the armed forces of the United States. Several million of these were young men scarcely eighteen years of age; of these, hundreds of thousands married childhood sweethearts before entering military service. These young wives and, often, their babies either waited at home for their husbands' return or occupied (largely substandard) housing near the stateside military bases where their husbands were stationed.

Let us assume (and I must confess disappointment that my sociologist colleagues have been unable to supply me with actual data) that a correlation exists between the marriage and childbearing patterns of mothers and daughters. That is, women who marry young and have their first child more or less immediately tend to have sons and daughters who in turn are receptive to the idea of early marriage and parenthood. This assumption is not implausible as far as different *groups* of persons are concerned; the marriage and childbearing patterns of the Irish and the Italians clearly differ. Whether within the same population the correlation between mother and daughter, or between parents and children generally, exists seems not to be known.

There need, of course, be no biology involved in the correlation that has been postulated above. The similar attitudes between parents and children within family lines could be accounted for (as far as I am concerned) by purely cultural factors—family traditions, for example, that are based on and fostered by family discussions and parental example.

That no biology is concerned does not mean that this cultural correlation is without effect on the population. Two segments of a population that reproduce at different rates—for whatever reason—change in their relative proportions: the more rapidly breeding one will eventually displace and replace the slower one. Now, the ratio of two exponential functions is itself an exponential function. In another of his essays, Hardin (1970) has said, "The social implications of this mathematical truth in a world that is finite, are radical in the extreme, and cannot be so much as mentioned without violating cherished taboos of our time."

Would the population of this example change only in cultural aspects? In mental aspects—attitudes toward sex and marriage, for example? Or in feelings toward togetherness or children's responsibilities and the like? Not at all! Consider our own population. More and more babies are being born to younger and younger mothers—many of these babies are illegitimate. According to news releases, one third of all births in both New York and Philadelphia are illegitimate. To the embarrassment of promotional groups in many cities, the first baby born after the start of the New Year is born to an unwed teenage mother.

Concurrently with the reports on the decreasing age of today's mothers, one also hears how poorly equipped physically many of these youngsters are for childbearing. This physical inadequacy brings us back once more to biology—to genetics, natural selection, and evolution. If in any population a long-lasting trend is set in motion which promotes childbearing at an early age, and if early adolescents vary in their physical ability to sustain pregnancies, then the population will change. Genes that tend to produce physical maturity at an early age will be favored within the population, and will increase in frequency at the expense of their alternative forms. The onset of the physical ability to reproduce easily will follow, with a short time lag, of course, the mean reproductive age of the population. Those who remember Aldous Huxley's *After Many a Summer Dies the Swan* may recall that the human species is a splinter of the higher primates in which reproduction occurs (in relative anatomical terms) at a preadolescent age. Many human anatomical features correspond to infantile ones of the great apes. That line of evolutionary change need not have ended, even now.

Conclusions

Many persons appear incredulous at the thought of natural selection; they look about (seemingly with closed eyes and minds), asking, "Where is the evidence?" There is no mystery about natural selection; examples are to be found everywhere: antibiotic-resistant strains of bacteria and other disease-causing organisms threaten the value of these substances in medical treatment; insect species develop resistance to insecticides more rapidly than chemists can discover new ones; unwanted plants develop resistance to herbicides, and mice and rats develop resistance to warfarin and other pesticides.

If within any population there exists genetically caused variation in the ability of individuals to survive a potentially lethal challenge (or to produce offspring successfully in the face of threats to reproduction), the characteristics of the population will change, perhaps gradually, as descendants of those who do survive and do reproduce come to constitute more and more of the population.

The most rapid and obvious changes in present-day human populations are those caused by man's cultural activities. The spread of the Christian and Muslim religions over much of the earth's surface represents a spread of ideas, not a diffusion of mutant genes. India's capacity to build and explode a nuclear device did not require an influx of genes from persons of Jewish, Italian, or Hungarian descent. Nevertheless, cultural evolution does not occur in a biological vacuum. Biological evolution evokes cultural change; cultural evolution evokes biological adaptation. Throughout most of human history, culture changed so slowly that biological adaptations occurred concomitantly. In today's world, the pace of cultural evolution has completely outstripped that of biological evolution. Adaptations to problems caused by today's culture must, for the most part, be cultural as well. The perfection of biological adaptations is a slow process. All artificial selection programs are accompanied by an initial reduction in the health and well-being ("fitness") of the selected plants or animals; considerable time is required for the selected population to recover from these associated disharmonies. The speed with which cultural changes occur may easily prevent the origin and development of mechanisms, even cultural mechanisms, that properly correct cultural disharmonies. Cultural evolution is not biological evolution, but an understanding of the fundamental causes and processes of one can help us anticipate problems (and possibly find their solutions) caused by the other.

CHAPTER 2

Natural Selection and Internal Teleology

No fewer than a million and a half species of animals, plants, and micro-organisms have been discovered and named. This is no more than one half, and possibly only one quarter, of all existing species. The remarkable thing about these species is the almost unimaginable variety of ways of life, often radically different, which they lead. There are autotrophs, which build their bodies of water, carbon dioxide, mineral salts, and solar energy. Hetero-trophs require organic compounds that have been synthesized by other or-ganisms; some are monophagous, specialized to feed on a single host or prey species or its products, and others are more or less polyphagous. Some live in salt water, others in fresh water, or in rainy and humid areas, or in arid deserts where the ability to withstand periodic desiccation is crucial. Some require warm temperatures, others resist arctic or antarctic colds; some need light, others live in permanent darkness; some are specialized predators or external or internal parasites, still others consume the dead or decomposing bodies of other organisms.

Descriptions of the ways living beings obtain their sustenance from the environment could fill several volumes. What interests us here is how it comes about that every living species is adapted to its particular way of life, different from that of every other species. Why is it that the body structures, physiological functions, and modes of behavior of every species enable it to survive and to reproduce in a certain range of environments in which it is usually found? Some species have mastered remarkably harsh and inhospit-able environments—hot springs, alkaline pools, rainless deserts, arctic wastes, and the like. Starting with Aristotle, biologists have marveled at the often astonishingly complex but effective adaptations of living beings to their respective environments. Living beings are teleological contrivances: They are built and function as if for a purpose. And this purpose is always the same: to maximize the chances for the survival of individuals, and for their leaving progeny.

Internal and External Teleology

Teleology and purposefulness are entirely absent in nonliving nature; they are universal in the living world. To talk of "purposes" or "adaptations" of stars, of mountains, or of physical laws makes no sense. However, artifacts devised by human beings, such as furniture or hammers or nails, although themselves nonliving, are made for certain purposes. Artifacts are contrived not only by humans but also by some animals. Human dwellings, and wasp or termite nests, are built for the purposes of shelter, food storage, or the housing and feeding of progeny. Nests and dwellings are nonliving, but their purposefulness (*telos*) is imposed by the living beings who construct them. Ayala (1968) has proposed that *external* and *internal* teleology be distinguished. An implement or a machine, such as a pencil or an automobile, is constructed by human beings to serve a certain goal, such as writing or transportation. Their structures and composition are imposed by external sources: their makers. Neither pencils nor automobiles ever build themselves. They are examples of external teleology.

By contrast, organisms, from the smallest bacterium to man, arise from similar organisms by ordered growth. They transport and transmit an impressive amount of ordering information. Their adaptedness, or internal teleology (or teleonomy), is contained within themselves. What is the origin of this internal teleology? It is a derivative of billions of years of evolutionary history.

Monod (1970, 1971) has stated these facts with admirable clarity:

Every artifact is a product made by a living being which through it expresses, in a particularly conspicuous manner, one of the fundamental characteristics common to all living beings without exception: that of being *objects endowed with a purpose or project*. . . . Rather than reject this idea (as certain biologists have tried to do) it is indispensable to recognize that it is essential to the very definition of living beings. We shall maintain that the latter are distinct from all other structures or systems present in the universe through this characteristic property, which we shall call *teleonomy*. (1971:9)

The origin of internal teleology is a fundamental, perhaps the most fundamental, problem of biology. Its proposed solutions fall neatly into two groups. One regards organic adaptedness, or internal teleology, as an intrinsic, immanent, irreducible, constitutive property of life. The other sees its source in the phenomenon of natural selection: internal teleology is not simply a static property; it waxes and wanes; its advances and recessions can be observed or inferred, sometimes induced experimentally, and their origin and causation analyzed like other biological functions.

Many authors have tried to evade the need to explain the origin of internal teleology by declaring it to be an elemental property of life. This evasion leads willy-nilly to explicit or implicit vitalism. Bernardin de Saint-Pierre (1784) ascribed the adaptedness of living beings to the generosity of Providence, who has arranged all things in the best possible ways. Archdeacon Paley (1802), in a book that was particularly influential in England until Darwin's time, argued that the perfection of living creatures proves the existence of a wise creator who contrived them so well. Immanent teleology is implicit in the pioneer evolutionist conception of Lamarck (1809, 1914). What else makes the heredity of organisms change in the direction of their needs and their efforts, rather than in random directions? Ridiculously, the same vitalistic assumption is basic to the allegedly materialistic, Marxist, dialectical, and other confusion of Lysenko, which was until a decade ago the approved orthodoxy in the Soviet Union.

Some otherwise great modern biologists have fallen into the same trap of assuming an elemental and inexorable purposefulness of living matter. It appears that a rejection of natural selection as the guiding agency of evolution makes this trap inescapable. In a most thoughtful (even brilliant) book, Grassé (1973, 1977) claims that the natural selection theory of evolution relies too much on chance (*hasard*). This way of exorcising chance merely invites immanent vitalistic teleology. Grassé makes this quite clear: "Immanent finality is an intrinsic property of all living creatures; without it they would not exist. . . . Immanent finality manifests itself in life itself, since this always tends to preserve itself and propagate" (1977:166–167). No wonder that the concluding sentence of Grassé's book is disappointingly pessimistic: "Perhaps in this area biology can go no farther: the rest is metaphysics" (1977:246).

What a penalty for having rejected the only rational explanation for internal teleology yet proposed!

Origin of the Idea of Natural Selection

Truly great ideas are sometimes remarkably simple—so much so that they occur independently to several persons at different times and places. What can be more obvious than that individuals who are hale and hearty have greater chances of surviving and becoming parents than do weak and sickly individuals? As Lewontin says, "in a world of finite resources, some organisms will make more efficient use of these resources in producing their progeny, and so will leave more descendants than their less efficient relatives" (1974:3).

It is that idea—so simple, so plain, so almost banal—that Darwin and A. R. Wallace made, in 1858, the foundation of their evolutionary theories. Nor is it unexpected that the idea of natural selection, of course not so named, occurred to several predecessors of Darwin and Wallace, starting with Empedocles (fifth century B.C.) and Lucretius Carus (first century B.C.). However interesting Darwin and Wallace's predecessors and anticipators may be, we must concentrate here on their successors. Darwin and Wallace regarded natural selection as the main propellant of evolutionary change. So do a majority of evolutionists today. Hence the rather inexact name "neo-Darwinism" is often used for the modern, biological, or synthetic theory of evolution. What are the outstanding developments of the theory of natural selection during more than a century since Darwin and Wallace?

Both Darwin and Wallace acknowledged that their insight into the operation of natural selection as an evolutionary factor was suggested to them by reading the work of the sociologist Malthus (1806), "An Essay on the Principle of Populations," which went through several editions between 1798 and 1826. Malthus postulated that human populations grow exponentially in number and therefore must inevitably outgrow the means of their subsistence. The growth is then arrested by starvation, disease, and war. Darwin and Wallace saw that the potentiality for exponential growth is quite universal in the living world. And yet populations of most species remain most of the time rather stable in numbers. Evidently only some of the progeny survive, while others are eliminated by death. Who survives and who dies is not entirely a matter of chance. In every population some individuals are stronger, or in some ways better adjusted to existing conditions, than are others. Their offspring are more likely to be represented in the population in the following generation than are the offspring of weaker individuals. Provided that vigor and other advantages are at least in part genetically conditioned, the incidence of these qualities will be enhanced in the subsequent generations.

Not even Darwin could avoid having his scientific findings subtly perverted by the sociopolitical forms of thought of his day. This was the golden age of laissez-faire, capitalist competition, and the building of colonial empires. What more appropriate than to describe natural selection as the outcome of a "struggle for life?" Darwin did point out that the world "struggle" is here used metaphorically. The struggle rarely takes the form of actual combat. Desert plants "struggle" against aridity by various devices guarding against excessive evaporation of water, not by extracting water from their neighbors. Among animals intraspecific struggle is often ritualized, thus avoiding actual harm to the participants. Nevertheless, "social Darwinists" (of whom Darwin was not one) saw in Darwin's theory a convenient justification for the oppression of the weaker by the stronger, and for race and class biases. "Nature red in tooth and claw" was the slogan that appealed to many.

Darwin accepted, with some hesitation, Herbert Spencer's description of natural selection as the "survival of the fittest." The use of the superlative exaggerates the fierceness of the supposed "struggle." In reality, what survives is not a single "fittest" but some of the tolerably fit. To some persons, however, "the fittest" evoked the image of Nietzsche's superman, a being who subjugates ordinary men. A rigorous definition of Darwinian fitness became possible only in our century with the development of population genetics. Anyway, Malthus' stress on death as a factor which eliminates excess population has yielded to an emphasis on differential birth rates. The surviving "fittest," if we still wish to use this phraseology, is no superman but rather the parent of the largest number of offspring.

In point of fact, models of natural selection can be constructed in which death plays little or no part. Suppose that in some generations the whole progeny produced by a population survives. Natural selection can still take place if the carriers of some genotypes are more fertile than those of others. This model may not be unrealistic. Some insects have two or more generations per year, and some generations (say, those hatched in the spring) are not limited by food or other resources. Here, selective values are determined by fertility. Or suppose that every pair of parents produces the same number of children (say, two), all of whom survive. If, however, the carriers of some genotypes marry and have children at a younger age than the others, they will have a higher selective value than those who marry late. Such a situation might well arise among human beings if for economic or other reasons (and with efficient birth control) all families were to be of the same size, but the age of marriage were to remain variable. Natural selection may still proceed under such conditions.

Mutations as the Raw Materials of Evolution

Darwin realized that some critically important parts of his theory could not, in his day, be fully elucidated. For example, what is the source of the variation on which natural selection operates? In 1909 the Danish geneticist Johannsen showed that the experimental selection he applied to pure lines of beans was without effect. The beans of his pure lines were still variable— some of the seeds were, for example, larger than others. But the average size of the offspring of the large and small beans from a pure line was the same, because the pure line was genotypically uniform. On the other hand, if one mixes several pure lines, selection does work: the progeny of large beans are on the average larger than those of small ones. In short, selection does not create genetic variation; it selects this or that kind of variant present in populations to which it is applied. For selection to work, the gene pool of a population must contain the raw materials, the genetic variation which can be selected.

Fortunately, we now know the source of the raw materials from which evolutionary changes are constructed. This source is mutation. We cannot take up here the matter of different kinds of mutation—genic and chromosomal, spontaneous and artificially induced. Suffice it to say that mutations occur in all organisms studied in this respect—from viruses and bacteria to higher plants and animals, including human beings. The time is not long past when mutations were regarded as rare, exceptional, cataclysmic events that could generate new species of organisms in single jumps. We now know that, although specific mutations of individual genes may indeed be rare, the aggregate frequency of mutations is quite high. Mukai (1964) found that in every generation as many as 14 percent of the second chromosomes of *Drosophila melanogaster* acquire new mutations. Clearly, within a few generations every chromosome will contain at least one mutant gene that was not present in it earlier. There is every reason to believe that human genes are, per generation, about as mutable as those of the fruit fly, *Drosophila*.

Of course, a great majority of the mutations in *Drosophila* have morphological or physiological effects so small that refined statistical methods are needed for their detection. Most, though certainly not all, mutations are due to the substitution of single nucleotides in the DNA of a given gene. This substitution often results, in turn, in the substitution of one amino acid for another in the protein specified by that gene. A single amino acid substitution may have a drastic phenotypic effect. Think, for example, of hemoglobin S, which differs from normal hemoglobin A by the substitution of valine for glutamic acid in one of 146 amino acid sites in the beta chain of the hemoglobin molecule. The gene coding for hemoglobin S causes, in homozygous condition, a sublethal disease—sickle cell anemia. Yet such a drastic effect is the exception rather than the rule. Probably a majority—how large a majority we do not know—of single amino acid substitutions cause subliminal phenotypic effects.

Obviously, mutations due to single amino acid substitutions do not generate new species. This was clearly recognized by Morgan as early as 1916. (There are mutations, those which double the chromosome complement of interspecific hybrids, which give rise to new allopolyploid species. We are forced to refrain in this book from discussing this interesting topic.) Most species, even closely related ones, usually differ in numerous gene changes. The best data concerning several species of *Drosophila* are those of Ayala and his colleagues (see Ayala et al. 1972). These authors have computed the genetic distances between geographic populations of a species and also between distinct species. Roughly speaking, genetic distance is an estimate of the proportion of genes coding for soluble proteins which have undergone at least one electrophoretically detectable amino acid substitution. Some characteristic genetic distances are:

Geographic populations	0.030 ± 0.006
Subspecies	0.228 ± 0.026
Sibling species	0.538 ± 0.049
Morphological species	1.214 ± 0.064

Sibling species are reproductively isolated but morphologically indistinguishable (or distinguishable only with difficulty); presumably they are closely related forms. Even *their* genes, however, have a sizable probability of having been changed by mutation. The objections raised by certain authors against the view that mutations are the building stones of evolution are demonstrably naive. All mutant strains of *Drosophila melanogaster*, we are told, still belong to *Drosophila melanogaster!* Species differences are compounded of hundreds, or thousands, or perhaps tens of thousands, of variant genes, ultimately arisen through mutational events. The process of mutation furnishes the raw materials for evolution; mutation itself is *not* evolution. The raw materials must become arranged into adaptively meaningful patterns. The "arranger" is the process of natural selection.

Mutation, Selection, and Chance

Fortuitous, random, luck, chance—all these epithets have been used to characterize the process of mutation. Because modern biological theory considers mutation to be the raw material of evolution, the same epithets have been applied by the critics of biological theory to the evolutionary process as well. Thus, it is imperative that some concepts and misconceptions be analyzed. No sane biologist ever ascribed evolution to "chance" or "luck" in the sense that these words are used by critics. Undeniably, there is an element of luck in biology, as there is everywhere in nature. But for evolution to generate internal teleology, chance must be subordinated to antichance agents. This subordination need not amount to a determinism so complete that the course of evolution be predestined from beginning to end (except in the sense of Laplacean universal deterministic causality). The interplay of chance and antichance leaves room for biological evolution to be a creative process.

Chance arises from the interplay of two independent causative forces (Nagel 1961; Delsol 1973). The substitution of one nucleotide for another in a segment of DNA is caused by events at the intramolecular and microphysical levels; such substitutions are probably the commonest source of gene mutations. We cannot predict which gene will mutate, or in which cell. In this sense mutation is said to be a matter of chance. This claim is no more than an indication of our ignorance. But several statements can be made about mutation which somewhat dispel ignorance. First, one can find out the frequency (that is, the probability) of mutations for a given gene. Thus, the mutation frequencies of some human genes are on the order of 10^{-5} per generation. Second, we know some physical (for example, X rays) and some chemical (such as nitrous acid and war-time mustard gases) agents which increase mutation frequency. Third, the kind of mutations that a given gene can undergo is determined by the structure of that gene, which in turn is determined by its evolutionary past. No gene can be transformed by a single mutation into just any other gene. The mutational repertoire of a gene may be great, but it is nevertheless limited. Thus, all mutations of the *white* locus in the X chromosome of *Drosophila melanogaster* have as their most obvious phenotypic effects various changes in the color of the fly's eyes.

Mutations can be said to be fortuitous in another, purely biological, sense. They arise regardless of whether they are useful or harmful where and when they arise. Mutations in species of bacteria conferring resistance to certain antibiotics arise whether or not the bacterial culture is exposed to those antibiotics. Mutations which confer resistance to DDT or other insecticides arise in house flies or mosquitoes that are not exposed to these insecticides. One can ask a naive but not unreasonable question: why did flies and mosquitoes have genes capable of making them resistant to insecticides before these insecticides were invented? Were they preadapted to resist insecticides? No, they were not preadapted *specifically* against insecticides. The genes in question have functions in insect metabolism, and it happens that some of these metabolic reactions detoxify certain insecticides. Another lucky event? Yes, if you wish. But do not conclude that flies have genes which can adapt them to any and all environmental insults. A race of flies capable of developing and reproducing at 50°C is unlikely to arise even under the most stringent selection program.

Selection does not operate on this or that gene, it operates on organisms which are endowed with this or that *constellation* of genes. This should be self-evident to the point of being platitudinous, yet some opponents of the biological theory of evolution have based their critiques on the notion that every gene is selected for (or against) independent of other genes. Even though each gene codes for only one polypeptide chain (a chain of many amino acids), the products of many genes interact during normal development. An individual is not the sum of its genes, it is an emergent product of its gene constellation. This fact is of basic importance in sexually reproducing forms.

According to Mendel's second law, an individual that is heterozygous for n genes is potentially capable of producing 2^n gametes with different gene constellations. A pair of parents each of whom is heterozygous for n different genes has the potentiality of giving rise to 4^n genotypes among their progeny. Even though chromosomal linkage makes some of these genotypes less probable than others, the theoretically possible gene combinations are far more numerous than the existing number of individuals of any species. How great are the values of n in natural populations? Studies on enzyme variation by means of electrophoresis have recently produced some data which permit a rough approximation in answering this question (see, for example, Ayala et al. 1972; Lewontin 1974). Two related measurements are made: first, the proportion of gene loci that are polymorphic, that is, represented by two or more alleles in a given population or a species; second, the estimated proportion of gene loci for which an average individual is heterozygous. The figures in Table 1 are extracted from Lewontin (1974).

The number of gene loci is not precisely (nor even approximately) known for any of the above organisms. They may number in the tens or hundreds of thousands, or even in the millions. For the sake of argument, let us take the minimal figure. Even so, the variety of possible gene constellations in each species is astronomically great. Consider human beings, who seem to be less polymorphic than, for example, many species of flies. One may still be quite certain that only a tiny part of all possible gene combinations has ever been, or ever will be, realized.

Table 1

Species	Percent polymorphic	Percent heterozygous
Homo sapiens	28	6.7 ± 1.8
Mus musculus	29	9.1 ± 2.3
Peromyscus polionatus	23	5.7 ± 1.4
Drosophila pseudoobscura	43	12.8 ± 4.1
Drosophila obscura	53	10.8 ± 3.0
Drosophila subobscura	47	7.6 ± 2.4
Drosophila willistoni	86	18.4 ± 3.2
Drosophila melanogaster	42	11.9 ± 3.7
Limulus polyphemus	25	6.1 ± 2.4

Let us not make the mistake of supposing, because all potential genotypes are merely combinations of a much smaller number of gene variants, that Mendelian recombination creates nothing new. It is *not* "plus ça change plus c'est la même chose." Ignoring monozygotic (identical) twins, there are as many human genotypes as there are persons living on earth. Admittedly, the individual characteristics of persons are largely products of their family, social, and cultural environments. Even so, human individuality is also genetically conditioned. Would anyone dare suppose that human individuality is inconsequential, and that the process which creates it creates nothing new?

Microevolution and Macroevolution

The most ancient Precambrian fossils are estimated to be some three billion years old (Schopf 1974; Barghoorn and Schopf 1966). They were probably all prokaryotes, apparently not unlike today's blue-green algae and bacteria. We assume that life on earth arose only once, and that all existing organisms are descendants of this primordial life. Biological evolution must then be said to be at least three billion years old. This much time has been taken in evolving higher plants and animals, including man.

Taking twenty-five years as the average generation for human beings in technologically developed countries, one generation is less than one hundred millionth of total evolutionary time. Suppose that a biologist studies evolution for thirty years. This time—a lifetime of research—is only one hundred millionth of the duration of evolution. If experiments could be carried out for, let us say, one tenth of all evolutionary time, a research biologist could make an ambitious attempt to reproduce under controlled conditions the evolutionary development of, for example, the vertebrate phylum. But this is out of the question. Experiments are, and probably always will be, confined to microevolutionary changes, taking place within time intervals very small relative to the known duration of evolution.

The terms micro- and macroevolution are also used in a different sense. The terms may indicate not duration of time but magnitude of change. By and large, the evolution of races, species, and perhaps genera is classed as microevolution.* The origin of families, orders, classes, and phyla is

---

* The term "microevolution" has acquired at least two meanings: the one given here, and that used by Dobzhansky (1954)—changes in Mendelian populations which are predictable on the basis of known adaptive values (see Wallace et al. 1983).

macroevolution. Experimental studies are limited to microevolution. Mendelian analysis is restricted to forms which can be hybridized and which produce fertile offspring. Comparison of amino acid sequences in homologous proteins, however, gives estimates of mutational distances, the numbers of mutations needed to transform the protein of one form into that of another. Such estimates can span classes and phyla.

The biological theory of evolution postulates that macroevolutionary changes are similar in origin and causation to microevolutionary ones. Opponents of the theory challenge the validity of this postulate. Can its validity be substantiated or falsified? To demand the experimental reproduction of macroevolutionary changes is clearly unreasonable. On the other hand, that we know only phenomena which cause microevolution is not conclusive proof that phenomena of a different kind do not remain to be discovered. Indeed, the opponents of the mutation-selection theory, from Goldschmidt (1940) to Grassé (1973, 1977), have without exception been driven to assume systemic mutations or the synthesis of new genes, for which no evidence whatever is available. The only reasonable course is to assume a substantial identity of micro- and macroevolutionary changes as a working hypothesis and then to see whether the hypothesis can be falsified by any well-established data.

First of all, we must state that the interaction of mutation and selection does produce adaptive changes on the microevolutionary level. We have already mentioned antibiotic-resistant strains of bacteria and insecticide-resistant races of insects. Such changes have been observed in many species and in many places, in "nature" as well as in laboratory experiments. There is no doubt that we are dealing with a widespread phenomenon. The resistant strains have had their adaptedness levels, their levels of internal teleology, enhanced with respect to environments that are contaminated with antibiotics or insecticides. To object that upon a return to uncontaminated environments some of these changes are reversed is not at all valid. Of course, the resistant strains belong to the same bacterial or insect species as the nonresistant ones. This shows only that species differ not in single mutant genes, but in many genes that have been altered by mutations.

Not all experimental evolution is confined to the infraspecific level. New species have been obtained by doubling the chromosome complement in hybrids between preexisting species (allopolyploidy). The classic example is *Raphanobrassica*, an allopolyploid obtained from a cross of radish (*Ra-*

phanus) and cabbage (*Brassica*). *Raphanobrassica* is fully fertile inter se, but gives sterile hybrids when outcrossed to either parent, radish or cabbage. (For more examples of species formation by allopolyploidy, see Grant 1963). True, this method of speciation is far more common in plants than in animals. Nevertheless, there is no doubt that it creates new species. Some experimentally obtained allopolyploids are similar to naturally existing species. Thus, one can experimentally recreate species that exist in nature or create more or less at will altogether new ones.

To produce or re-produce species without allopolyploidy is certainly difficult, because, as stated above, numerous genetic differences would have to be combined; yet the task is not entirely hopeless. The superspecies *Drosophila paulistorum* is composed of six incipient species. The incipient species seem to be identical morphologically, but they cross only with difficulty (that is, they are reproductively isolated), and among the hybrids produced, males are entirely sterile (Dobzhansky and Boesiger, 1968). One strain, derived from a single female found in Llanos, Colombia, has become, after several years of cultivation in the laboratory, incapable of producing fertile male hybrids with the incipient species with which it was formerly fertile. By means of artificial selection for some fifty generations, this strain also became sexually isolated from strains with which it formerly crossed easily (Dobzhansky et al. 1976). This strain may also be regarded as a new incipient species.

Selection and the Origin of Complex Organs

Every biologist, classic or modern, who is acquainted with living beings cannot fail to admire the purposeful complexity of the structures and functions of these beings. This complexity is to be seen everywhere, literally everywhere, in both the so-called simple and higher organisms. Structural complexity and functional perfection reach almost unimaginable heights in such organs as the human brain and the vertebrate eye. Can such sublimely perfect organs have arisen and evolved through the interaction of mutation and selection?

Darwin himself was troubled by the question of whether natural selection can account for the complexity of vertebrate eyes. Many evolutionists since Darwin doubted or denied that it could. They claimed that organs such as the human eye and brain are instances not of internal teleology generated by natural selection but of external teleology imposed by divine will through mysterious guiding forces. They insisted that the evolutionary process is finalized (that is, its eventual outcome is foreseen at the outset), while natural selection is a chain of chance events, unable to give rise to purposeful complexity.

The ability of natural selection to construct complex organs is a test of the power of the biological, or synthetic, theory of evolution to explain macroevolutionary changes. The test is difficult to make because, as pointed out above, the applicability of experimental methods to the study of macroevolution is severely limited. The battle of arguments must inevitably be fought on the field chosen by the finalists, or even by antievolutionists. Nevertheless, a self-consistent model, consonant with the known facts of modern evolutionary genetics, can be constructed.

A human eye consists of numerous parts, coadapted in the sense of Cuenot (1941), or even of Darwin (1859). For an eye to see, it must have a pupil, lens, iris, cornea, crystalline body, retina with its rods and cones, and the proper connections with the brain, all simultaneously present at their respective locations. The absence or abnormality of any one of these parts blinds the eye, or at least sharply lowers its visual acuity and precision. It is unbelievable that mutations producing every one of these parts occurred simultaneously. But, if they did not, the future eye was useless to its possessor. Natural selection, far from maintaining such a useless organ, should have gotten rid of it. It has done so in many cave animals which are blind but are descendants of ancestors with functioning eyes. The opponents of selection theory assume that eyes were premeditated by some agency directing the evolutionary process and that this agency managed to get all the proper eye parts assembled at about the same time.

The above account contains several fallacies. First of all, it is not true that the human eye, or the eye of an eagle, appeared in an ancestor which had empty spaces in the places now occupied by eyes. Our vertebrate ancestors had eyes capable of vision, but eyes different from our own. In the cyclostome fishes, we do find eyes built more simply than ours. The lancelet (*Amphioxus*) is not entirely deprived of vision, but it has only photosensitive pigment cells in the anterior portion of its neural tube. So do some of the tunicates, which are probably related to the ancestors of the vertebrates. Natural selection did not build eyes ex nihilo. It gradually improved eyes, or photosensitive cell clusters, which were not as complex and not as efficient as our eyes but were nevertheless useful to their possessors (Salvini-Plawen and Mayr 1977).

The human brain is an even more complex organ than the eye. It is probably the most powerful, though not the most voluminous, brain known in any animal. It contains some ten billion neurons that perform different specialized functions and that are interconnected with one another or with sense organs in the most intricate and functionally effective ways. It would be absurd to suppose that mutations for every neuron and every synapse must have come together for the brain to begin to function. Comparative anatomy shows a remarkable sequence of brain forms: brains become smaller and less complex as we descend from man to other mammals, reptiles, amphibia, and fishes, to tunicates, amphioxus, and the chaetognaths. The vertebrates are supposed to have had common ancestors with the echinoderms. These animals have a nervous system that is constructed quite differently from those of the vertebrates. Natural selection did not create the human brain ex nihilo; the human brain is a product representing a gradual improvement of brains of other kinds.

Another fallacy, more often implicit than expressly stated in many discussions of the origins of complex organs, is the assumption that there exist genes for each component part of the organ. The preformationist idea that each gene represents in the sex cell a "unit character" or a body part was entertained by some pioneer geneticists early in the current century. It was abandoned at least a half century ago in favor of a more epigenetic view. There are no separate genes for the pupil, the iris, the lens, and every other anatomical part of the eye. The eye is a product of a developmental process visibly starting in the early embryo; it is an integral part of the coordinated events whose totality transforms a fertilized egg into an infant and eventually an adult body. To say that all of an organism's genes must participate in the development of the eye might be an exaggeration. The point is simply that there is no one to one correspondence between a gene and an anatomically delimited body or organ part. (This is true even though some hereditary diseases show that a mutational change in a gene may manifest itself primarily in one "target organ." Such seemingly limited expression merely shows that the gene normally present at the altered locus is *necessary* for the normal development of that organ; in no sense does it demonstrate that the unaltered gene normally present at that locus is *sufficient* to give rise to a normally developed organ.) Nor are there special genes determining the characteristics of a class or an order separate from those determining the characteristics of a genus or a species.

To what extent evolutionary progress has depended on an increase in gene number and in the amount of "genetic information" is a complex and still unresolved problem. True, the DNA content of prokaryotes is much smaller than that of eukaryotes. But it is not true that the so-called higher or more advanced organisms always have more DNA per nucleus than the less complex ones. Among vertebrates, it is not man or mammals but some amphibians that have the most DNA. Progressive evolution among vertebrates appears to take place by changing genes rather than by adding new ones. This is not to say that there are no known mechanisms by which gene number can be increased. One such mechanism is gene duplication; another is polyploidy. Both duplication and polyploidy cause certain gene loci to be represented more times than formerly. The duplicated genes may then undergo different mutational changes and acquire functions which the ancestral genes did not have. Yet it is a completely open question whether a given species, for example, *Homo sapiens*, has some gene loci not present in, say, the chimpanzee, or even in a mouse. What is certain is only that man has numerous gene loci represented by alleles differing from those in other primate or mammalian species. Species differences are compounded of many gene variants.

Mutations are far more numerous, and on the average less drastic, than supposed by classical geneticists. The speed of evolutionary change does not depend on mutation frequencies. The situation is far more complex and interesting. Mutational raw materials are probably always available. The old idea that the evolutionary conservatism of species that failed to evolve for long geological periods reflected an absence of mutations is certainly wrong. Selander et al. (1970), in studying genetic variability in the horseshoe crab, *Limulus polyphemus*, showed that this "living fossil" has levels of genetic polymorphism and heterozygosity comparable to those of *Drosophila* or man.

Rapid evolution depends on environmental challenges being responded to by natural selection. The position of a species in the ecosystem and its interrelations with other organisms and with its physical environment decide whether or not it evolves and, if it does, in what direction. Much research is still needed in this field; we are only beginning to understand what can serve as an evolutionary challenge, and how it can be responded to. In some situations the action of natural selection can be compared to a sieve. A single gene mutation can make a species of bacteria resistant to an antibiotic, or a species of insect to an insecticide. In an environment containing this antibiotic or this insecticide, natural selection does act as a sieve: all sensitive individuals die, and only the resistant ones survive (and multiply, thus amplifying their numbers).

It would be naive to envisage all evolution, particularly that of complex organs, according to the sieve model. For example, the evolution of vertebrate eyes and brains did not consist of the appearance of rare mutations, one of which fabricated the lens, another the iris, a third the cornea, and so on for the other parts of the eye. The process can be better envisaged if we suppose that the gene pools of the populations of ancestral (as well as those of now living species) always contained an abundance of genetic variants modulating developmental processes. The variants that were selectively favored at any moment were unlikely to have resulted in drastic anatomical changes. Most probably they were polygenes, each one of which had only slight modifying effects. Drastic changes do occur, of course. This is demonstrated by the existence of numerous genetic malformations and diseases recorded in every carefully studied species, particularly in man. Such changes are selected against by normalizing selection. They are the constituents of the species' genetic load. Polygenic variations are more difficult to isolate and to study singly, but there is no doubt that they are present in the gene pool of every species. They are responsible for the individual variations among "normal" persons in the species *Homo sapiens* and among the "normal" individuals of other species, as well.

Conclusions

We do not claim that the problem of the origin of adaptations, of internal teleology, is resolved. Far from it! We have here a field demanding much further study. It is necessary, for example, to harmonize the discoveries of molecular geneticists with those of the ecological, experimental, and mathematical geneticists. A really synthetic theory of evolution has yet to be created. Discoveries of fundamental importance may follow from such a synthesis.

In this chapter, we have argued that the working hypothesis which should guide us toward the new synthesis is that of the modern version of natural selection. Nothing yet discovered in biology contradicts natural selection. The origin and evolution of complex organs is a severe test of natural selection theory, particularly because we are here deprived of the possibility of validating our conclusions by appropriate experiments. Macroevolutionary changes took many millions of years to run their courses; it is ridiculous to demand that they be reproduced in the laboratory.

Still, the finalistic, orthogenetic, and neo-Lamarckian theories which have been advanced as alternatives to evolution by natural selection are in far worse states. They drive their proponents into blind alleys of two kinds. Neo-Lamarckian theories assume that characters acquired in the somatic parts during an individual's lifetime are passed to its progeny by inheritance. This assumption has consistently failed its critical tests. Molecular genetics has shown that such inheritance is not merely improbable but impossible. Finalistic and orthogenetic theories postulate that evolution is made directional by mysterious agents or forces. No such agents have yet been discovered; this failure forces the partisans of such theories to appeal to future discoveries, which they can only hope somebody will make sometime. This is an unenviable position for a scientist.

CHAPTER 3

A Critique of Non-Darwinian Theories of Evolution

We have endeavored to demonstrate in chapter 2 that the biological, or synthetic, theory gives a self-consistent explanation of the basic phenomena of biological evolution. Internal teleology and the adaptedness of organisms to their environments are no longer mysteries suggesting the action of inscrutable forces. They can be understood as the products of natural selection. Nevertheless, we have admitted that far from all evolutionary problems have been resolved. The idea of natural selection, inherited from Darwin, has been greatly developed and supported by a multitude of facts. In its application to many evolutionary problems, however, it still remains a working hypothesis that needs to be critically tested. Theories other than the synthetic one (often, and inaccurately, labeled "neo-Darwinian") have been suggested as alternative explanations of evolution. It is our intent to consider critically some of these supposed alternatives. Even if none of them are found to be valid, their analysis may disclose aspects of evolutionary phenomena requiring modifications or extensions of the synthetic theory.

Potentiality and Preformation

The universe is the product of an evolutionary process. All that was, is, and will be has evolved, is evolving, and will evolve. Past evolution has brought about the present, and the present is the foundation of future evolution. If this is so, then the potentialities of both present and future states must have existed in the past. Primordial life must have had the potentiality of evolving mankind, as well as every one of the several million now living and extinct biological species. The statement that the potentiality of man was contained in the primeval living monad, and in the primordial cosmic stuff, may seem inordinately audacious. In reality, the statement is trite! This affirmation does not consider that primordial life had, in addition to those potentialities that were realized, countless unrealized ones.

If earth were larger or smaller than it is, or warmer or colder, or had a larger or smaller portion of its surface covered with water, or mountains, or plains, it might have evolved quite different organisms, or life might have been extinguished soon after its origin. Neither mankind nor any other existing species was predetermined to arise, except under the exact circumstances that actually existed at all stages of its history. What other forms of life could have evolved under other circumstances, we do not know. Modern biological knowledge is much too inadequate for such inferences. Some scientists, mostly cosmologists and chemists rather than biologists, make bold statements that extraterrestrial life, which they assert *must* exist, *must* have developed rather similarly to life on earth. This assertion includes the evolution of rational life, the so-called extraterrestrial humanoids. These bold assertions lack support in present biological knowledge.

Leeuwenhoek, the pioneer Dutch microscopist, discovered near the end of the seventeenth century the sperm cells of animals and man. An egregious error was soon added to this discovery. Some microscopists imagined that they saw inside the head of the human spermatozoon a homunculus, a tiny human figure. This error led to a theory of preformation. If a sex cell contains a miniature copy of an infant or of an adult organism, then embryonic development—ontogeny—is essentially growth. The homunculus increases in size, becoming an infant and eventually an adult man or woman. If ontogeny is preformed, so, logically, could be phylogeny. Etymologically, evolution means unfolding or unrolling something that was invisibly present in ancestral forms. The term "evolution," before Darwin, was used in describing embryonic development. This prior use makes understandable Darwin's reluctance to use the word "evolution" in *On the Origin of Species.*

The alternative to preformation is epigenesis. Epigenesis means the development of something that did not previously exist. Like preformation, epigenesis may be used in describing ontogeny and phylogeny. Of course, eighteenth-century ideas of preformation and epigenesis seem quite naive today. Not so, however, the philosophic conceptions underlying these ideas.

Both individual and evolutionary development may be conceived as realizations of programs contained in genetic material. Epigenesis does not mean creation ex nihilo. A fertilized human ovum does not contain a homunculus, but neither is it a structureless drop of viscous liquid. It contains, in addition to nutrient materials, a developmental program encoded in the DNA of its chromosomes and cytoplasmic particles. Ontogeny—individual development—is the realization of this program, modulated by environmental conditions. Does the DNA also contain a program for phylogenetic development? Almost surely, no.

To guard against misunderstanding, the above concepts may be reiterated in a different way. In any organism, ontogeny is so programmed that an ovum or a bud or a bulb yields either an individual of a certain species or nothing (that is, it dies before completing development). Even so, programming is not absolutely inflexible; in different environments development proceeds in subtly or conspicuously diverse ways. Thus, human development depends, to use somewhat antiquated words, both on nature and on nurture. *Evolution lacks a program.* This is not to say that phylogeny is wholly unconstrained or wholly at the mercy of the environment. A mouse is unlikely to evolve into a species of elephant, or vice versa. Their organizations are so distinct that they could hardly be transformed into each other, although they have descended from a common ancestor in the distant past. Nevertheless, a species confronted with an environmental challenge may respond by adaptive genetic changes. "May," rather than "must;" the response depends on many factors, such as the availability of genetic variation on which natural selection can work. Neither biological nor human evolution arises wholly from within the organism or wholly from the environment. Each involves creative syntheses of eternal and external causations.

Autogenesis and Directed Evolution

The program encoded on the DNA of a given species gives rise to orderly and directed embryonic and postembryonic development. This development, especially in higher organisms, proceeds through a series of complex and marvelously orderly transformations. Within certain limits, its end is determined in advance. This end is the adult organism, capable of producing sex cells which will give rise to a new cycle of development. Time and again, many authors—especially in the late nineteenth and early twentieth centuries—were seduced into believing that evolution is, in principle, similarly programmed. Evolution is, they believed, autogenetic. Its program is contained within the organism; evolutionary changes are predetermined, orderly, and directed. Ontogeny occurs within an individual's lifetime; autogenesis occurs gradually, over a long series of generations.

There are not one but several variant autogenetic theories of evolution, often differing radically from one another. Some are frankly vitalistic, others less frankly so. Some postulate that evolution is guided by supernatural forces extrinsic to the organism. Others say that evolution is nomogenetic (Berg 1926), determined by unknown "laws" residing in living matter. Some postulate rectilinear orthogenesis and a direct (but predetermined) course arising within the organism itself. Others (Grassé 1973, 1977) do not insist on rectilinearity. To Osborn (1917, 1921) evolution was aristogenetic, driven toward betterment and perfection.

We neither can nor need to analyze autogenetic theories one by one. All of them end by postulating an internal finality as an intrinsic property of living beings. The analogy between ontogeny and phylogeny is misleading. The program of ontogenesis is a product of phylogenetic development, not vice versa. The finality, the internal teleology, of the individual organism is, of course, unquestionable. This is what any biologist familiar with living beings is constantly impressed, even awed, by. Yet, as we showed in chapter 2, internal teleology is what any theory of evolution must explain in order to be acceptable. The theory of natural selection does precisely that, and in our opinion does so successfully. To assume, as theories of autogenesis do, explicitly or implicitly, that this teleology is immanent in all living beings means to take for granted what is to be explained. It is a *petitio principii* of a most flagrant kind.

The most recent (and thoroughly argued) autogenetic theory is that of Grassé (1973, 1977). Thanks to his incomparably comprehensive knowledge of the living world, Grassé had marshaled an impressive abundance of paleontological and comparative anatomical evidence. The evolutionary changes which he discussed cannot, Grassé says, arise from chance, from random mutations. But he is like Don Quixote fighting windmills. No one claims that mutations by themselves make evolution. Grassé's notions of mutation and natural selection are much too simplistic. There is no one-to-one correspondence between genes and mutations on the one hand and organs and body parts on the other. The idea that each gene represents a "unit character" of the body was entertained by some pioneer geneticists but, with the exception of some behavioral geneticists, has been abandoned for more than half a century. As we showed in chapter 2, the genetic basis of a complex organ, such as the human eye, is not a mosaic of genes each of which is responsible for one constituent part.

It is nearer the truth to say that the entire genotype induces the developmental pattern of the body and that the eye (with its individual parts) is an integral part of this pattern. Genes produce enzymes and proteins, not body parts. Mutations are common, not rare; genetic variance is seldom in short supply. Natural selection is not a sieve which mechanically retains some mutations and discards others. Because the number of different genotypes, in sexual species, nearly equals the number of individuals themselves, natural selection has an abundance of material on which to act. In acting, it behaves more like an engineer or a builder than like an expurgator or destroyer. The emergence of amphibious from fishlike ancestors, and of mammals from reptilian progenitors, did not occur because some extraordinary mutations happened to arise. These emergences were rather the results of a constructive action of natural selection, repatterning and coordinating the times of action of many genes, and requiring long periods of time.

Grassé is even less successful when he passes from the critical to the affirmative part of his book. If mutation and natural selection cannot bring about evolution, where does evolution come from? It comes, according to Grassé, from the addition of completely new genes which miraculously advance evolutionary progress. Grassé (1977:228) is forced to admit that "no formation of new genes has been observed by any biologist, yet without it evolution becomes inexplicable. . . ." This statement is incorrect. New genes are formed through the duplication of old genes, followed by a divergence of both the structure and the function of the duplicated genes by means of an accumulation of mutations. Thus, higher vertebrates have separate genes for several kinds of hemoglobins and for myoglobins; these are duplication products of a single ancestral gene. Grassé ignores such facts because to admit them would mean acceptance of the evolutionary role of mutations in natural selection. He declares, "To make no distinction between the mutation and the creation of a gene is truly to understand nothing of the innermost mechanisms of evolution" (Grassé 1977:230–31).

Alas, the creation of a gene de novo, not by the modification of a preexisting gene, is a piece of fantasy. All genes of all organisms are descendants of a primordial gene that arose three or four billion years ago in primeval life. De novo origin of a gene would amount to de novo origin of life! Suppose that some enzyme would string along free nucleotides into double helices. Even so, what miracles could make these chance combinations function as useful genes, as bearers of evolutionary progress? It is not surprising, then, that the autogenesis postulated by Grassé, like all other autogenetic theories of evolution, ends by an appeal to internal finality. We read: "To eradicate finality from biology is a vain attempt because it goes against reality . . ." (Grassé 1977:129). One should distinguish more carefully between internal finality and internal teleology. The former implies vitalism; the latter finds its explanation in natural selection.

Neo-Lamarckism

Lamarck was the author of the first self-consistent theory of the evolution of the living world. His place among the greatest biologists who ever lived is completely secure. But it is not surprising that, more than a century and a half ago, Lamarck could not give an account of evolutionary mechanisms that is acceptable today. His basic idea of "gradations" makes him the author of the first theory of autogenesis. According to Lamarck, the graduations stem from an intrinsic drive of life toward perfection. They result in a gradual but inevitable progress of living beings, from the simplest to the most complex.

Another factor that causes evolutionary changes is the influence of the environment. Among animals with developed nervous systems, this influence is indirect. Changed environments change the needs of their inhabitants, and changing needs change habits. Some organs are exercised and strengthened, while others are used less and are weakened. The results of the use and disuse of organs are, in the Lamarckian view, inherited. The inheritance of acquired modifications was, in Lamarck's time, so nearly self-evident that there was little need to prove that such inheritance really occurred. Even half a century later, Darwin still accepted the inheritance of acquired traits, although as only a subsidiary factor of evolutionary change. It was the supposed inheritance of acquired characters, of course, that became associated with Lamarck's name and was regarded by neo-Lamarckians as the principal, if not the only, factor in evolution. The heyday of the neo-Lamarckians was the late nineteenth and early twentieth centuries. The evolutionary role of natural selection was denied, or at least underestimated; natural selection was believed, at best, to be the guardian of the purity of the species, sweeping away deleterious changes. Evolutionary changes were induced by the environment, either by direct modification of body characters or indirectly, by the use or disuse of organs.

August Weismann in 1883 challenged the belief that characters acquired during an individual's lifetime can be transmitted to its offspring (Weismann 1892, 1893). His belief in the "omnipotence of natural selection" made him the founder of neo-Darwinism. Using the name "neo-Darwinism" for the modern biological theory of evolution is at best inexact and at worst misleading. Let us admit, however, that at the turn of the century, both neo-Lamarckism and neo-Darwinism were reasonable and legitimate working hypotheses. Numerous experiments that were designed to prove or disprove the inheritance of acquired traits were made and published. Among the last of such experiments were those of Kammerer (1924), which caused heated polemics during the 1920s. Since then, except for the abnormal recrudescence of Lysenko in the USSR, neo-Lamarckians have gone into eclipse; Lamarckism has few if any competent defenders today.

It is quite wrong to lump neo-Lamarckism with the autogenesis of Lamarck and with autogenetic theories, as is commonly done. They are really polar opposites. Neo-Lamarckism is an ectogenetic rather than an endogenetic theory. Some persons, mostly nonbiologists, still ask, "How can we be sure that acquired characters cannot be inherited?" After all, our experiments extend at best for a few years—some dozens or (for microorganisms) hundreds of generations; nature has at its disposal millennia, and countless generations. What we now know about the molecular basis of heredity shows, however, that numbers of years or of generations are irrelevant. A gene is a segment of DNA which encodes the sequence of amino acids in a protein molecule. Because of its structure, a gene makes a true copy of itself. To be sure, the process of copying is at times inexact, so that different nucleotides are sometimes substituted into the gene and different amino acids into the protein. We also know that mutation frequencies, the rates at which copying errors are made, are enhanced by shortwave radiation and by certain chemicals (mutagens), such as the mustard gases used in World War I and many modern industrial compounds.

How can a change in the body that is caused by the use or disuse of organs (or by anything else) become reflected, mirrored, and copied in an appropriate change in the nucleotide sequences of one or more genes? *Equifinality* is the crux of the matter; not just any change will do. Suppose that exposure to sunlight causes an increase of skin pigmentation. There are genes some of whose alleles code for enzymes that generate more and other alleles for enzymes that generate less pigment. But there is no known method whereby more pigment in the skin can induce nucleotide substitutions in these genes such that new, mutant alleles will be produced which code for enzymes that would induce more pigment in individuals of the following generation. Note this argument well: *the hypothesis that acquired characters are inherited demands not simply that mutations be induced in genes by changes in the body, but that these mutations be directional ones causing equifinal effects in the progeny.* If at the turn of the century the neo-Lamarckian hypothesis seemed plausible, it has now become so farfetched as to be discarded as worthless.

Neutral Traits and Neutral Mutations

A useful genetic variant (by definition) increases the selective value, the Darwinian fitness, of its carriers. Natural selection will tend to increase its frequency in the population. By contrast, a deleterious variant (also by definition) decreases Darwinian fitness. Natural selection operates to diminish the incidence in populations of carriers of deleterious traits. But, as stated explicitly by Darwin himself, natural selection will neither encourage nor counteract the spread of inherited variants that neither increase nor decrease fitness. Metaphorically, one can say that natural selection is "blind" with respect to neutral traits.

Suppose we examine the traits differentiating races of a species, or species of a genus. Are these traits useful to their possessors? This may seem to be a simple way to test the validity of the theory of natural selection as a guide to evolutionary changes. The test seems to lead to negative results; the adaptive significance of most racial and specific characteristics are not at all obvious. This is particularly true if one examines taxonomic keys, used by systematists for determining species. For example, some species of *Drosophila* have divergent and others convergent scutellar bristles. Why should it be better to have these bristles diverge or converge? It was the lack of obvious adaptive significance of the differences between races and species that led, at the turn of the century, to a disenchantment with the theory of evolution by natural selection. For lack of something better, some anthropologists, zoologists, and botanists declared that systematics should be based on neutral characteristics, which allegedly reflect phylogeny and are not perturbed by adaptation to the environment.

The problem of seemingly neutral trait differences remains unresolved even at present. One possible solution is that these differences are not really neutral. A visible morphological trait, such as convergent or divergent scutellar bristles in species of *Drosophila*, may be only an external manifestation of adaptively significant physiological differences. In other words, convergent or divergent, scutellar bristles need not be useful of themselves, but they may reflect pleiotropic effects of gene differences that also produce quite other effects. One should never forget that there is no one-to-one relation between genes and "unit" characters.

It has been shown in chapter 2 that biological species differ in numerous genes, not in just a few, as classical geneticists thought. Genes do not act independently of one another, each gene for itself. They act in concert; the genotype is not a mosaic of autonomous units, but an organized and integrated system. A gene that produces an adaptively irrelevant external effect may be important as a constituent part of the genetic system of the species. That species' genotypes are coherent genetic systems is attested by, among other things, the well-known facts of partial or complete inviability or sterility of many species hybrids. This includes hybrids between sibling species of *Drosophilia*, the morphological differences between which are small and seemingly of no adaptive consequence.

Nevertheless, it would be needlessly dogmatic to insist that any and all genetic changes arising by mutation must be either useful or harmful and that all racial and specific differences are adaptive. Darwinian fitness is a continuous variable; the effects of genetic change range over the entire spectrum, from lethality, through neutrality, to advantageousness. Genetic changes resulting from mutation, selection, and migration can be deterministic provided that the rates of mutation, selection, and migration are known and are constant over time, and that the population is infinitely large. One can then predict by rather simple calculations the genetic composition of any future generation. But in small populations consisting of only some tens or hundreds of individuals, stochastic changes occur as the result of sampling variations (chance errors) in a gene pool of limited size. Stochastic changes in small populations are called random changes, random drift, or genetic drift. Random drift will be most noticeable with adaptively neutral, or nearly neutral, genetic variants. In the case of favorable or deleterious variants, natural selection tends to counteract drift.

Random genetic drift has been studied for many years in a series of theoretical papers, especially in his classic work of 1931, by Sewall Wright. We need not go into the mathematics describing drift under various conditions. Wright, it may be noted, never claimed that genetic drift alone produces important evolutionary changes. Rather, the *interaction* of drift with deterministic forces is important. Many animal and plant species are subdivided into numerous colonies, some of which may consist of rather small numbers of individuals. If colonies do not exchange migrants, or do so rarely, they will diverge with time in gene frequencies. The population sizes of the colonies, and the environments in which they live, will probably not be constant. Natural selection may periodically be relaxed or become more severe. Wright believes that isolated colonies can act as "scouting parties" exploring the field of gene combinations. A species that is subdivided into numerous colonies may evolve more effectively than one large, continuous species.

During the 1930s and 1940s, there was a considerable misunderstanding of the intent of Wright's genetic drift theory. Some biologists, mostly nongeneticists, believed that drift could account for evolution independent of natural selection. In particular, because many racial and specific traits do not obviously have an adaptive significance, their emergence in evolution might be ascribed to the allegedly non-Darwinian factor of genetic drift. Mathematical intricacies, inevitable in the analysis of stochastic evolutionary changes, helped this misunderstanding. A reaction came in the 1950s. Some genetic variants previously believed to be neutral were shown to be under the control of natural selection. Although the population sizes of some colonies of many species are undoubtedly small, these colonies are not always rigorously isolated from each other. Now, it had already been shown by Wright himself, in the thirties, that very small exchanges of migrants between colonies suffice to obliterate the genetic divergence expected on the basis of random genetic drift. Most evolutionists in the late 1950s entertained a point of view that was described, half in jest, as "naive panselectionism." During the 1960s, the genetic drift hypothesis came back with redoubled force, amounting in some instances to a "naive panneutralism." The reasons for this revival must now be considered.

Is the Genetic Load Too Heavy?

We know (see chapter 1) that useful mutations are rare. Most mutations that arise in any organism are deleterious or, at best, quasi-neutral. Many are lethal. The pressure of new mutations is counteracted by normalizing natural selection. Under a stable environment, an equilibrium of mutation and selection pressures is reached at which the number of deleterious mutations arising per unit time is equaled by the number eliminated by selection during the same time. The equilibrium values for deleterious mutations at each locus are described by simple formulas: for example, $q = \sqrt{u/s}$ for recessive ones (q = equilibrium gene frequency, u = mutation rate, and s = intensity of selection). The mutations still to be eliminated constitute the mutational load. How heavy is this load?

Already in 1950, Muller realized that in man as many as 30 percent of newly formed sex cells may carry a newly arisen mutation at some locus or another. New studies suggest that this figure is an underestimate. Now, if one assumes (as Muller did) that almost all mutations are more or less deleterious, one faces a paradox. Because, at equilibrium, the number of mutations must be equaled by the number of selective eliminations ("genetic deaths," in Muller's terminology), the mutational genetic load alone seems too heavy for any species to carry. It should be noted that the number of genetic deaths caused by dominant deleterious mutants is even greater than the mutation rate, because dominant genes are eliminated mostly in heterozygous carriers, carriers who have but *one* mutant allele. It should also be remembered that the rate of elimination at equilibrium is equal to or greater than the mutation rate whether the mutation produces a drastically deleterious effect or only a slight loss of fitness.

The paradox becomes even more awkward when one considers the problem of polymorphism and its maintenance in natural populations. Some genes are represented in human populations, or in populations of other sexual species, by two or more alleles. If an allele is rare, say 1 percent of the gene pool ($q = 0.01$), it may plausibly be supposed to be maintained by mutation pressure opposed by natural selection. But this mechanism is unlikely for more frequent alleles, say where $q = 0.10$. Mutation rates that would have to be assumed in order to maintain such high frequencies are unrealistically high. Consider, for example, the blood group polymorphisms of human populations. At the Rhesus (Rh) locus in man there are at least twenty alleles, of which at least five reach frequencies of 0.10 or more in certain populations.

There are several forms of balancing natural selection that can maintain the polymorphisms found in many species. One of the simplest (although not the only one possible) is heterotic balancing selection (see chapter 2). If the heterozygote, A_1A_2, has a fitness greater than both homozygotes, A_1A_1 and A_2A_2, then both alleles, A_1 and A_2, will persist in the population. But a heterotic balancing selection imposes a "genetic load" on the population. Assume that the heterozygotes, A_1A_2, have a fitness of 1.00 and that both homozygotes, A_1A_1 and A_2A_2, have fitnesses 10 percent lower (0.90). When the population reaches an equilibrium, the frequency of the heterozygotes will be 0.50 (50 percent) and that of each homozygote 0.25 (25 percent A_1A_1; 25 percent A_2A_2). We can easily calculate the average fitness of the population: $0.50 \times 1.00 + 0.25 \times 0.90 + 0.25 \times 0.90 = 0.95$. The average fitness will thus be 5 percent less than the maximum fitness, 1.00, which could theoretically be found in a population consisting entirely of heterozygous (A_1A_2) individuals. In reality, as shown in chapter 2, in a sexual, cross-fertilizing species, a population composed only of heterozygotes would inevitably produce 50 percent of homozygotes in the next and every subsequent generation.

A loss of 5 percent of the maximum possible fitness might still be tolerated in most biological species, even if this means the "genetic death" of 5 percent of all progeny produced. A polymorphic population would have the advantage of including many individuals of the fittest heterozygous (A_1A_2) genotype. The question that arises, however, is this: How many heterotic balanced polymorphisms can a population afford to maintain? Suppose that a population has 100 balanced polymorphisms, each reducing the population fitness 5 percent below the theoretical maximum; that is, each having a mean fitness of 0.95 instead of 1.00. Make the additional simplifying (but probably unrealistic) assumption that the polymorphisms and their effects are independent of each other. The mean fitness of a population with 100 polymorphisms would then be 0.95^{100}, or only 0.006—about six tenths of 1 percent of the maximum fitness. This is a very low average fitness indeed. The "cost" of maintaining 100 balanced polymorphisms is, on the assumptions we have made, probably too high for any species to "pay." The reproductive potential of the species would be reduced by more than 99 percent.

Classical and Balance Models of the Population's Genetic Structure

We seem to be in danger of getting into a cul-de-sac. The genetic loads which populations carry seem to be insupportable. A possible escape is to suppose that balanced heterotic polymorphisms are rare and exceptional phenomena in most species. This is the escape that Muller (1950 and later works) and other adherents of the classical model of population structure tried to take. The intellectual roots of this theoretical model go back to the ideas of such pioneer geneticists as Morgan, de Vries, and Bateson. Briefly, the model assumes that most individuals of most species are homozygous for "normal," or "wild type," alleles at a vast majority of all gene loci. Thus, each population or race of mankind consists of individuals homozygous for the "normal" genes of that population or race. Another race (or another species) would likewise be homozygous for those genes that are "normal" for it. Mutation generates abnormal mutant alleles of low fitness at some gene loci. Their spread is counteracted by normalizing natural selection, so that their frequencies are kept low. Balanced polymorphisms are temporary expedients: if one exists, a mutation will eventually arise which will have a fitness, in its homozygous condition, equal to or surpassing that which was previously possessed by the then superior heterozygotes. This new favorable mutation will eliminate the polymorphism. So, according to Muller and others, genetic variation in populations is merely a necessary evil.

The balance model was developed particularly by Lerner (1954) and Wallace (1968, 1981) and discussed by Dobzhansky (1970) and Lewontin (1974). It postulates that in populations of sexually reproducing and outbreeding species, most individuals are heterozygous at many, perhaps at most, gene loci. Since healthy and "normal" individuals are heterozygotes, there is no "normal," or "wild type," allele at most gene loci. It is, of course, admitted that alleles of low fitness produced by mutation are present at low frequencies in the gene pool. What is important, however, is that balanced polymorphisms of many genes are common. Far from being a necessary evil, genetic variation is an adaptive mechanism with whose aid the species masters its many environments.

To contrast the quasi-philosophical implications of the classical and balance models is interesting. The ideal human population, or the best adapted population of any species, would, according to the classical model, consist of individuals homozygous for optimal alleles of all genes. One should strive to locate, to identify, or to select such an optimal genotype, and to promote its spread until it becomes the "type" of the species. The balance model emphasizes the value of genetic diversity both for the normal development of the individual and for the generation of diversity among individuals. An ideal human population would include a variety of genotypes, none of them unconditionally superior to others in respect to health or fitness, but most efficient in their diverse ways under different circumstances. If, in addition, the best-adapted genotypes are heterozygous—and perhaps heterotic—the polymorphisms are essential to the species. The implications of the classical and balance models lead, among other things, to different programs for eugenics. That topic, however, is outside the framework of this chapter.

For several decades there was considerable dispute among evolutionary geneticists as to whether the classical or the balance model best describes population structures found in reality. The dispute was not settled but shifted to a more sophisticated level in 1966 when Lewontin and Hubby published their studies on enzyme polymorphisms in natural populations of *Drosophila pseudoobscura* (see Lewontin 1974 for a review of developments in this field). Previous to this work, geneticists could count polymorphic genes but could not record the number of monomorphic ones. Consequently, they could not estimate the proportion of gene loci which are polymorphic in natural populations. The technique of electrophoresis of water-soluble enzymes has changed the situation. The enzymes to be studied are chosen not because genetic variants have previously been found but because chemists have developed staining procedures for their detection. The proportion of polymorphic loci was found to be far greater than the adherents of the classical model of population structure ever thought possible. Table 1 (chapter 2) summarizes some of the data. In animals ranging from man and other mammals to species of *Drosophila* and *Limulus polyphemus*, from 23 to 86 percent of gene loci studied were found to be polymorphic. An average, "normal" individual is heterozygous for 5.7 to 18.4 percent of its genes.

These proportions are patent underestimates: only those variants of protein structure can be detected by electrophoresis which change the electric charge of the molecule, and such changes are only about one third of the changes which protein molecules undergo. There is no doubt that many loci recorded by even these sophisticated techniques as monomorphic are, in fact, polymorphic.

Whether enzyme-coding genes are more or less variable than other genes is still an open question. As a working hypothesis, one can assume that the genes studied thus far are not a grossly distorted sample of the entire genotype. If so, a human individual must be heterozygous for some tens of thousands, and human populations polymorphic for hundreds of thousands, of genes. This is obviously incompatible with the wild type concept, the supposition that variable genes are maintained in natural populations by mutation pressure. On the other hand, it is difficult to understand how vast numbers of polymorphisms, particularly those detected by electrophoretic techniques, can be maintained. If they are maintained by heterotic balancing selection, the genetic load (granting the simplifying assumption of independent—that is, multiplicative—action mentioned above) is simply too heavy for any population to bear. Evolutionary genetics seems to have reached an impasse.

The Attempt to Cut the Gordian Knot

Several authors, particularly Kimura and Crow (1964), Kimura and Ohta (1971), and King and Jukes (1969) have made a heroic attempt to escape from the impasse. The last-named authors gave their theory the name "non-Darwinian evolution." This name is inexact; several theories of evolution discussed in this chapter, beginning with that of Lamarck, were also "non-Darwinian," because they did not consider natural selection to be the chief propellant of evolutionary change. The escape is to assume that most polymorphisms are neither adaptive nor deleterious, but simply neutral. Neutral genetic variants impose no genetic load. They are neither promoted nor discriminated against by natural selection; natural selection is, as we have said earlier, "blind" to neutral traits.

The proponents of the "non-Darwinian" school claim that most mutations on the molecular level are neutral. Variant proteins detected by electrophoresis differ from each other most often by the substitution of a single amino acid, a change that alters the electrical charge of the molecule. Such a substitution need not affect the enzymatic or other functions of the protein, unless it happens in a part of the molecule that is of key importance for these functions. Some parts of protein molecules are free to undergo changes, while others are conserved. The neutralist school admits that some mutations have deleterious effects; this is too obvious to be denied. A small minority may even be useful. What is necessary is to dispose of the theoretical difficulty (the seemingly enormous genetic load) which arises not only if all but even if a substantial fraction of *existing* polymorphisms are *not* adaptively neutral.

If polymorphisms arise from adaptively neutral mutations, one must explain what makes any neutral allele increase in frequency, or even displace an earlier allele. The answer is random genetic drift. Random drift is, of course, more important in small than in large populations. Consider a diploid population of a genetically effective size N. The genetically effective size is usually smaller, often much smaller, than the number of individuals of which the population is composed. Suppose that a gene in this population has 2N alleles, no two of them identical, but all neutral with respect to Darwinian fitness. In the following generations, most of these alleles will be lost by chance, but others will multiply (also by chance) and be represented in many individuals. Given enough generations of random drift, only a single allele that was present in the original population will remain. Kimura and Ohta (1971) deduce mathematically that the chance of any one of the 2N original alleles reaching fixation is 1/2N, and that the average number of generations required until fixation is 4N. They deduce further that, in the long run, the number of neutral alleles reaching fixation per generation will be (on the average) equal to the number of new alleles arising by mutation (2Nu).

Evolutionary Clocks

The neutralist theory is ingenious, and works well—at least on paper. In recent years important studies have been made comparing the sequences of amino acids in homologous proteins of different organisms. The hemoglobins and the cytochromes c have been studied most extensively, but the number of additional proteins "sequenced" is rapidly increasing. As could be expected, the amino acid sequences are more similar in phylogenetically closely related forms and less so in phylogenetically remote organisms. Nevertheless, cytochromes c still preserve similarities in the most diverse organisms, from man and mammals to insects, plants, yeasts, and molds. Techniques have been developed for estimating the "mutational distances" between the proteins of different organisms. Mutational distance is measured as the minimum number of mutations that must be assumed in order to account for the divergence of the amino acid sequences of homologous proteins. The mutational distance between the cytochromes c of man and monkey (*Macaca*) is 1, of man and rabbit 12, and of man and yeast (*Saccharomyces*) 56.

Partisans of the neutralist theory make two bold assumptions: first, that mutations producing amino acid substitutions in proteins are (and have been in the past) adaptively neutral, and second, that the rates of fixation of such mutations are uniform and proportional to time. If these assumptions are granted, by knowing the mutational distances between species of organisms, one can estimate the time that has elapsed since these species descended from their common ancestor. The estimate obtained can be checked against the time estimates obtained by paleontologists from studies on fossil organisms.

The estimates obtained from molecular evolutionary clocks and those based on paleontological information are generally in fair agreement. However, we ourselves do not believe that this agreement bears out either the assumption of the neutrality of most mutations or of uniform rates of fixation through time. Evolutionary rates may be uniform on the average over long time periods, and this independently of whether amino acid substitution is due to selection pressure or to genetic drift. Alternation of periods of fast and slow evolution will go undetected because of averaging. Attempts to read the "clock" for relatively short periods have led to contradictory results: Wilson and Sarich (1969) have inferred that the divergence of the hominal and pongial primates took place only four or five million years ago, but paleontological evidence gives a much longer time estimate—twenty to thirty million years.

Different proteins evolve at sharply unequal rates. According to King and Jukes (1969), the rate of amino acid substitution in mammalian fibrinopeptide A is about four times greater than that in the alpha and beta chains of hemoglobin and about thirteen times greater than that in cytochrome c. To save the idea of biological clocks, one must assume that some amino acid locations in a protein are essential for its function and therefore few or no substitutions at these locations are tolerated, while other locations are less restrictive and at some the amino acids can be freely changed. This is another way of saying that natural selection conserves certain amino acids in fixed places in a protein molecule, while mutations and changes at other locations are more easily accepted. Some enzymes, for example, cytochrome c, act similarly in a test tube no matter from what organism they are derived. To take this as evidence of their interchangeability in living organisms, however, seems rather naive to a biologist.

Race and Species Differences in Protein Polymorphisms

In some organisms, protein polymorphisms vary in frequency in different geographical populations of the same species. Human beings and *Peromyscus* (mice) are among the examples of this geographic variation (Cavalli-Sforza and Bodmer 1971; Selander et al. 1971). On the neutralist theory, this difference is interpreted as meaning that gene frequencies have drifted to different values in geographically different but conspecific populations. The same data are, of course, also compatible with the hypothesis that natural selection has made certain alleles more and others less frequent in accordance with the environments prevailing in different geographic regions. A contrary situation has been found in other organisms. Several species of *Drosophila* have shown great similarities in the enzyme polymorphisms present in different geographic populations of the same species, even in populations living at great distances from each other (see Lewontin 1974; Ayala et al. 1974 for reviews). For example, *D. willistoni* has a distribution area extending from Florida to the La Plata river in Argentina. Yet, with few exceptions, the same loci are polymorphic and are represented by approximately the same allele frequencies everywhere. How is such astonishing similarity to be explained? One possibility, favored by Ayala and his collaborators, is that the polymorphisms are maintained by heterotic balancing selection and that the relative Darwinian fitnesses of the heterozygotes and homozygotes are uniform in the habitats in which the species lives.

The neutralist school claims that the above data favor their position—that the enzyme polymorphisms are neutral. If so, however, why does genetic drift fail to induce differentiation of the geographic populations? Because, they say, the whole species is effectively a single population; the uniformity is maintained by migration, and as we have seen above, very little migration suffices to cancel the genetic divergence of populations with respect to selectively neutral alleles as a result of random drift. Very small rates of migration cannot be detected experimentally, some *Drosophila* individuals may on rare occasions be transported passively by wind (or other means) for considerable distances. Nevertheless, most biologists are staggered by the idea that a species like D. *willistoni*, likely numbering billions or even trillions of individuals and spread over territories some of which are many thousands of kilometers apart, can be effectively a single population!

Ayala et al. (1974) have tried to analyze the quantitative predictions of the neutralist theory by testing them against their observations on enzyme polymorphisms in species of *Drosophila*. Kimura and Crow (1964) and Kimura and Ohta (1971) have derived a prediction for the numbers of electrophoretically distinguishable neutral alleles per gene locus to be found in populations. This number is $n = \sqrt{1 + 8Nu}$, where N is the genetically effective population size and u is the mutation rate per locus per generation. Using the lowest plausible estimates, $N = 10^9$ and $u = 10^{-7}$, one would expect to find some twenty-eight different alleles per locus. The number n of alleles found experimentally is about 1.215, or twenty-three times fewer than predicted.

Other predictions of the neutralist theory also disagree with observations (Ayala et al. 1974). Among the approximately thirty gene loci coding for electrophoretically detectable enzymes in *Drosophila* species of the *willistoni* group, some are more and others less polymorphic. In natural populations, heterozygotes for the more variable loci are more frequent than for the less variable ones. The average heterozygosity, $\overline{H}$, is 0.177. On the assumption that the polymorphisms are neutral, the frequency of heterozygotes for most loci should cluster about the average value, while deviations from the average should be rare. In reality, the most frequent class, about one quarter of all gene loci, is monomorphic with no heterozygosity ($H = 0$). The observed values of H for different loci range from 0 to 68 percent.

The neutralist hypothesis leads one to expect that the configurations of allelic frequencies will be different in different species. Because species do not exchange genes through migration, there is no reason for the frequencies of electrophoretically detectable alleles at corresponding loci to be uniform. In reality, different species such as *D. willistoni*, *D. equinoxialis*, *D. tropicalis*, and *D. paulistorum* have similar allele frequencies at many loci, very different frequencies at other loci, and intermediate situations at only a minority of loci (Ayala et al. 1974). Thus, some alleles are diagnostic for a species, having frequencies approaching unity in one species and zero in another. Other alleles have, on the contrary, approximately the same frequencies in two, three, or all four species. Least frequent are intermediate degrees of genetic similarity. This situation might be expected if the frequencies of alleles are under the control of natural selection, but not if the alleles are neutral.

Thus far, the only report of enzyme polymorphisms which differ from one geographic population of a *Drosophila* species to another is that of Johnson (1971) for *D. ananassae* and *D. nasuta* from islands of the Pacific Ocean, stretching from Samoa in the east to the Philippines in the west. Several enzyme-coding gene loci have allele frequencies that differ from one island group to another. This finding is compatible with either the neutralist or the selectionist hypothesis.

Correlations Between Enzyme Polymorphisms and Environments

Several attempts have been made to discover a possible effect on Darwinian fitness of different alleles coding for electrophoretically variant enzymes. The most successful of these are the studies of Clegg and Allard (1972) and Hamrick and Allard (1972) on a species of wild oat, *Avena barbata*, in California. In the part of California characterized by an arid climate, this species is monomorphic, while in the more humid parts it is polymorphic for five enzyme-coding gene loci. Where these loci are polymorphic, a correlation is found between allele frequencies and the aridity of the habitat. Different allele frequencies are observed in populations growing only some hundreds of meters apart. The genetic composition of the plants in these dry microhabitats is like that in the geographic areas where the species is monomorphic.

Equally convincing are the results of Koehn and Rasmussen (1967) in the fish *Catastomus clarkii* in the Colorado River basin in the western United States. The frequencies of certain alleles vary greatly, correlating with the temperature prevailing in a given locality. This correlation has also been confirmed in laboratory experiments on this species of fish.

Dobzhansky and Ayala (1973) found indications of a seasonal change in the frequencies of certain enzyme alleles in a natural population of D. *pseudoobscura*. Such seasonal changes have been known for a long time for polymorphic chromosomal inversions in the same and other species. Powell (1971) and McDonald and Ayala (1974) made experimental populations of D. *pseudoobscura* of which some were deliberately kept in variable and others in relatively constant environments. The founders of these laboratory populations were samples of natural populations which were polymorphic for numerous enzyme alleles. The polymorphisms were conserved in the variable environments but were reduced in the constant ones. This is, once more, what one would expect if enzyme polymorphisms are maintained by natural selection.

Conclusions

It goes without saying that attempts to find adaptive correlations of ostensibly neutral genetic variants have been made for only a tiny minority of such variants. This means neither that most such variants are really neutral nor that their effects on Darwinian fitness remain to be discovered. Some genetic variants which were for a long time believed to be neutral were eventually shown not to be so. Associations between blood groups and such diseases as duodenal and stomach ulcers, cancer of the stomach, and pernicious anemia have been established. Claims of associations with plague, smallpox, and syphilis have also been made, but not confirmed. That the associations thus far established explain observed geographic variation in blood group frequencies is, however, doubtful; such associations do not in themselves prove the nonneutrality of these alleles (see a review in Cavalli-Sforza and Bodmer 1971).

Some points must be remembered in the quest for selectional explanations for polymorphisms. First, the fitness differences between genotypes may be too small to be detected without an inordinate amount of work. A fitness advantage for heterozygotes amounting to 10 percent, not to speak of 5 percent or 1 percent, over their corresponding homozygotes would rarely be detected in either natural or experimental populations of higher organisms.

Nevertheless, such fitness differentials are amply sufficient to maintain observed polymorphisms. Second, differences in fitness may exist in some environments and not in others. Thus, chromosomal inversion homo- and heterozygotes in D. *pseudoobscura* differ greatly in fitness at 25°C, but at 15°C no differences have been demonstrated (reviewed by Dobzhansky 1970). A polymorphism may be adaptive during the breeding season but not during hibernation, or vice versa. The dependence of fitness on the environment may be particularly relevant for mankind. It is evident that the human species exists now in environments quite different from those of its ancestors of a million years ago, or even of ten thousand or one thousand years ago. Quite possibly, then, some of the genetic variation in human populations represents phylogenetic relics which were adaptive in the past but are neutral today.

It would be dogmatic to deny that some genetic variants may indeed be neutral. Because of the degeneracy of the genetic code, some nucleotide substitutions in DNA do not alter the amino acids in the proteins encoded by the genes in which such substitutions occur. Here we have mutations, by definition, but mutations that are undetectable by protein analysis. Can such mutations influence fitness? The only possibility of such an influence would be through transfer RNAs. Lest we forget, the Darwinian fitness of a genotype is a continuous variable. The carriers of one genotype may have a slight advantage over the carriers of another in one environment, a disadvantage in a second, and equal fitness in a third.

The "non-Darwinian" neutralist hypothesis represents an attempt to salvage what remains of the classical hypothesis of population structure. Lewontin (1974) calls it the "neoclassical" hypothesis. We do not think this name is justified, because the essence of the classical model was the assumption that most individuals of a race or a species are geneotypically nearly identical and homozygous at most gene loci. That contention is no longer supported by any knowledgeable person. The neutralist model acknowledges the hitherto unsuspected genetic variety in populations but regards most of this variation as genetic noise of no adaptive significance. The emergence of the neutralist model has been for many years, and continues to be, a stimulus for research on and thought about evolutionary problems. Whether this model is eventually confirmed or denied, it has nevertheless played a useful role in biological science.

CHAPTER 4

The Unique Position of the Human Species in Evolution

Renan claimed that the apex of intellectual evolution is to understand man. This is, of course, no more than a variation on a more ancient theme: know thyself. Scientists should not pretend that science alone is enough to understand man. What is certain, however, is that understanding man without science is impossible. Anthropology is the science of man. But the science of man is not anthropology alone—man is the subject of study of all biological and all other sciences. In brief, all science is anthropocentric. We study the world of living things and of inanimate nature, the planet Earth, and even the remotest galaxies in order to understand man and man's place in the cosmos. There are more than two million species of organisms now living on earth. Humankind, *Homo sapiens*, is only one of them. Nevertheless, science devotes more attention to this single species than to all others combined. This is as it should be. To study man alone, however, and to neglect all other species would be counterproductive for understanding ourselves and others of our species.

The generally accepted view during the Middle Ages was that the world was created by God expressly for man. Strictly speaking, science can neither prove nor disprove this view. It became less credible, however, when Copernicus, Galileo, Kepler, Newton, and others discovered that the universe is almost unimaginably vast. The planet Earth is like a speck of dust in cosmic space. The vastness of the universe and the smallness of man were frightening to Pascal. Man no longer seemed an essential part, let alone the center and purpose of the universe.

Descartes made a gallant attempt to restore man's primary status. Animal bodies are mechanical automata; they lack consciousness and souls. Human bodies are also automata; they do, however, have souls, every individual his own. The possession of a soul sets man apart from and above the rest of nature. Cartesian dualism has proved to be unconvincing. The evidence for human souls is not compelling, but if one admits that human beings do have souls, why deny them to other beings? One of the schools of philosophy and theology presently fashionable in the United States is *process philosophy*, inspired particularly by A. N. Whitehead. Process philosophers believe that there are rudiments of life and consciousness not only in all living things but also in molecules, atoms, and subatomic particles. Even this panpsychism, however, does not restore the Cartesian dualism of soulful man and soulless brutes (Birch 1974; Rensch 1974).

Humankind: A Product of Evolution

When Linnaeus, in the mid-eighteenth century, classified all living beings known at his time, he placed the species *Homo sapiens* with the apes in the zoological order Anthropomorpha. To Linnaeus, this did not mean that humans and apes were descended from a common ancestor. That they were so descended seems to be precisely the conclusion drawn by Lamarck (1809, 1914), although (because of the political situation of his time) he felt obliged to put his conclusion under disguise. Here is one of several characteristic statements he made: "If man was only distinguished from the animals by his organisation, it could easily be shown that his special characters are all due to long-standing changes in his activities and in the habits which he has adopted and which have become peculiar to the individuals of his species" (1914:169–170).

Darwin, in his great work *On the Origin of Species* (1859), refrained, for reasons similar to Lamarck's, from an explicit discussion of man's evolutionary origins. Only in *The Descent of Man* (1871) did Darwin set forth his conclusions: man is a part of nature, kin to all that lives, and his nearest relatives among recent organisms are apes and monkeys. Everyone knows what a storm of protest Darwin's publications unleashed. Traditionalists maintained that man's dignity had its foundation in the belief that he appeared in 4004 B.C., essentially in his present state. Although enlightened theologians saw the groundlessness of the fears that evolution destroys human dignity, biologists and anthropologists had the duty to validate beyond all reasonable doubt the theory of evolution, especially as it applies to man. Nothing less than overwhelming evidence on this matter could suffice.

The efforts of evolutionists have succeeded thoroughly. Already by the turn of the century, no informed person could reasonably doubt that mankind evolved from ancestors who were not men. At present only ignorant persons and fanatics are antievolutionists. (Alas, they have a political influence that is not negligible in California and elsewhere within the United States.) The fact of evolution as a historical happening having been thoroughly established, the attention of evolutionary anthropologists and biologists could turn in another direction. The classics of evolutionarism emphasized the many respects in which human beings are fundamentally similar to other biological species. At present, it is most important to study in what ways they are *un*like other species.

The biological uniqueness of mankind is at present the central interest of human evolutionists. The situation has been well described by Julian Huxley:

Man's opinion of his own position in relation to the rest of the animals has swung pendulum-wise between too great and too little a conceit of himself, fixing now too large a gap between himself and the animals, now too small. . . . The gap between man and animals was here reduced not by exaggerating the human qualities of animals, but by minimizing the human qualities of men. Of late years, however, a new tendency has become apparent. (1941:1–2)

This new tendency is the recognition that man "is another species of animal, but not just another animal. He is unique in peculiar and extraordinarily significant ways" (Simpson 1964:24).

Morphological Differences

From the time of Linnaeus to the present, the classification of organisms has been based on differences in their bodily structures. The human species is morphologically clearly distinct from any other species, including its nearest living relatives, the apes. The magnitude of these distinctions is great enough to place mankind in a monotypic genus *Homo* of the monotypic family *Hominidae*. The several living species of apes constitute the family *Pongidae*. But the major differences between humans and other animals belong to the realm not of morphology but of behavior. If the classification of animals were based on psychological capacities rather than on body structures, the evaluation given to such differences could well be that man's uniqueness deserves a rank greater than the familial one.

Man's morphological peculiarities have to do principally with two aspects: an erect posture and the possession of a brain larger in relation to body size than those of the other primates. Man is not the only bipedal animal. For example, kangaroos move by leaping on their hind legs, and so do some rodents. But man is the sole species that walks comfortably with the trunk erect. The posterior extremities have become specialized for locomotion, while the anterior ones are able both to carry objects and to perform delicate operations. Natural selection has evidently favored this division of labor in the human ecological niche. And vice versa, man's dependence on the manufacture and use of tools was promoted by selection because of his bipedal stance. It would be wrong to suppose that either a dependence on tools or a bipedal stance came first: they coevolved in a cybernetic relationship—natural selection promoted by mutual feedback. This is apparently a common phenomenon in evolution. One may also see here a modern version of Lamarck's notion that bodily changes are induced by needs and by the behavior of organisms. This notion need only be corrected by recognizing that behavior and needs are altered by natural selection through corresponding bodily changes.

The human bipedal stance required numerous changes in the bones of the pelvis and in the musculature connecting the pelvis with the trunk and legs. To support the skull with its heavy brain on top of the vertical (more precisely, S-shaped) spinal column, the *foramen magnum* and the neck joint moved forward to the middle of the bottom of the skull. Some of the necessary changes brought biological disharmonies to man's bodily structure. The bowl-shaped pelvis must support the weight of the abdominal viscera and must also provide the opening through which the infant passes at birth. Although the female's pelvis is somewhat wider than that of the male, the infant's head, with its voluminous brain, is barely able to squeeze through. Thus, the process of childbirth is often agonizingly painful.

The difficulty of childbirth has resulted in another change which from one point of view can be regarded as a defect. Human infants are born at a less advanced developmental stage than are other primate infants. Infant apes and monkeys soon after birth are able to hold firmly onto their mothers' bodies; newborn human infants are utterly helpless. They have to be carried by the mother, thus making her less mobile than she might otherwise be. To be sure, the prolonged childhood and delayed sexual maturity have been turned to good use in the human species. It gives time for the process of socialization of children, a process that is indispensable for a human being to become a functional member of society. One could, of course, extend the list of defects and imperfections of the human body.

Is the existence of these defects compatible with the theory of evolution by natural selection? Should not natural selection have made everything physically flawless? The answer is NO! Natural selection, as Jacob (1977) has expressed it, is a tinkerer, not an engineer. It does not operate with this or that gene separately from the others. What survives or dies, produces progeny or fails to do so, are not genes or even organs. Survival and successful reproduction are events that befall individuals—or groups of individuals known as Mendelian populations. A defect in one organ, provided that it is not completely lethal, can be compensated (even overcompensated) for by the efficiency of other organs. What counts in the balance is the total efficiency of the organism's ability to perpetuate itself and the genes responsible for its development. Despite difficult and painful childbirth, the human species is not extinct. Its biological success is the fruit of its brain power, not of its body power.

Natural selection leads to the adaptation of the human species (or any other species) to the present. It builds on the past; it knows nothing of the future. Of course, the opportunism of natural selection is its Achilles' heel. The commonest end of an evolutionary line is extinction. The great majority of Mesozoic species, and an even greater majority of Paleozoic species have left no living descendants. How is it that natural selection permits extinction? One reason is that a species adapted to one environment may be genetically unable to reconstruct itself rapidly enough when that environment changes. Mankind is a species that is genetically capable of originating culture. It has, in fact, done so, and in so doing it has become completely dependent on cultural environments.

Would the human species become extinct if culture were suddenly erased? Or could it readapt itself to conditions that existed for its prehuman ancestors of, say, ten million years ago? Prophesies in this field are impossible.

They are impossible because mankind, if it so wishes, can direct its evolution on any course of its choice—including deliberate extinction. We cannot foresee what direction our descendants a century or a millennium—let alone a million years—from now will wish to take. Certainly, mankind is the only biological species that ever existed that is aware of its origin by evolution and that has the potential ability to direct its future evolution. None of the million species which became extinct in the past could do anything to prevent extinction. If confronted by a danger to their continuation, human beings could take steps to avert the danger. But will they do so? History so far is not completely reassuring on this score. Barring a cosmic catastrophe, the extinction of mankind would be the first instance of evolutionary suicide of a species. At present, human beings possess the technological means for committing species suicide. Misuse of nuclear energy may be the instrument. One can only hope that they will have the collective good sense to avoid such misuse.

Culture

At the close of the Pliocene, or early in the Pleistocene, there appeared an extraordinary animal species that began more and more to rely on learned rather than on genetically fixed behavior in securing its livelihood. This species began to evolve culture. It was in this ancestor of Homo sapiens that culture became the paramount adaptive mechanism. Culture is the totality of information and of behavioral patterns that are transmitted from individual to individual, and from generation to generation, by instruction and learning, and by example and imitation. Certainly, culture depends ultimately on the human genetic endowment; nonhuman animals at most have minute rudiments of cultural transmission. What human genes transmit is the potentiality for the acquisition of culture, not culture itself. Culture must be acquired by every individual for himself. There are no genes for the acquisition of French, of Chinese, or of Australian aboriginal culture. Every nonpathological individual can, especially during childhood, acquire any of the many existing human cultures. Here is an analogy: Human genes are indispensable for learning human languages, but they do not determine which of the many existing languages a person will learn, let alone what the person will choose to say in that language.

Possession of a genetically assured potentiality for the acquisition of culture and language makes *Homo sapiens* biologically unique. It also makes our species adaptively the most successful product of organic evolutionary history. With no little arrogance man has called himself the lord of creation. Examples can easily illustrate the superior power of cultural over genetic adaptation. Birds, bats, and some other animals are more or less proficient fliers. They are all descendants of nonflying ancestors. Their evolution involved gene changes which induced the development of variously constructed wings. Man, however, has become the most powerful flier of all, and without transforming limbs into wings. Instead, he builds flying machines. His body did not acquire wings; his brain constructed them.

That adaptation by cultural change is more important to man than adaptation by genetic change is incontrovertible. New inventions are more significant than new mutations. Changed genes are transmitted only to direct descendants of the individuals within which the changes arise. Many generations of selection are needed in order to confer the benefits of the changed genes on the whole species. Changed ideas, skills, or inventions can be transmitted, in principle, to any number of persons within a single generation—regardless of biological descent—through writing, printing, radio transmission, and other techniques of communication. The propagation of cultural changes may occur regardless of the separation of teachers and pupils in either space or time.

Language

The potentiality for the transmission of culture is due to genes. The realization of this potentiality depends mostly on human symbolic languages. One must distinguish between human languages and so-called animal languages. Communication by means of visual, acoustic, and chemical signals is virtually universal in the animal world. Human mental activity is characterized by preoccupation not with signals but with symbols. Human languages are predominantly symbolic. A symbol is an act, or an object, the meaning of which is socially agreed upon or bestowed by those who perform this act or utilize this object. Most words in any language bear no resemblance to what they symbolize; their meanings are learned. The signs used in animal communication, for example, the barking or tail-wagging of a dog, is understood by another dog with little or no learning.

We do not know just when in evolution mental activity based on symbols began. At any rate, as Monod has stated it: "An impartial observer, let's say someone from Mars, could not fail to be struck by the fact that the development of man's specific performance, symbolic language—a unique occurrence in the biosphere—opened the way for another evolution, creator of a new kingdom: that of culture, of ideas, of knowledge" (1971:128).

We would only emphasize that this uniquely human and superorganic evolution has a genetic foundation that has been constructed by natural selection during the biological evolution of our ancestors. Understand us well: we do not mean that the genetic foundation was completed before there was a widespread use of symbols. The genetic capacity to make and use symbols and the eventual utilization of these symbols have mutually reinforced each other. They developed by a process of positive feedback; each became more and more evolved as the other became strengthened and perfected. As with many pairs of traits, the capacity to use symbols and their actual use coevolved.

The recognition of the evolutionary uniqueness of man as a user of symbols does not mean that species other than man are absolutely devoid of the capacity to use them. An evolutionist who finds an unusual characteristic in one species looks in related species for materials from which this characteristic can be constructed. Several investigators have made experiments trying to teach animals to use at least rudiments of human symbolic language. Chimpanzees, seemingly our nearest living relatives, have been a favorite experimental subject. Most successful were the experiments by Premack (1971). He used colored plastic figures as symbols. A young chimpanzee learned first the use of the symbols for objects, then for qualities such as colors and shapes, and for abstractions such as "similar" and "different." With a vocabulary of several dozen symbols, the animal was able to understand and to use "words" and finally learned to combine them into simple "sentences." Gardner and Gardner (1969) also succeeded in teaching a young chimpanzee a symbolic language consisting of gestures, originally devised for communication among and with deaf-mute persons. Least successful were the older experiments of Hayes and Hayes (1954), who attempted to teach a chimpanzee ordinary auditory symbols. The uses of symbols can, at least to a limited extent, be taught to chimpanzees; that much has been proven. What use, if any, these animals might make of their symbolic capacity in the wild state is uncertain.

Self-Awareness and Death Awareness

The one truth that Descartes found beyond doubt is "cogito ergo sum"—
I think, therefore I am. My being aware of my thought is unquestionable
evidence of my existence. My awareness of myself, my esprit, started, as far
as I can tell, in my early childhood. Except in sleep or under narcosis, it
has never left me, and it will probably remain with me until death. My
mind, my self-awareness, is to me the most immediate and indubitable cer-
titude. However, how do I know that other persons also possess minds and
self-awareness of their own? I cannot possibly enter their selves and experience their sense impressions, feelings, or thoughts. I can only assume that
they probably have an awareness of themselves, as I have of myself. The
evidence in favor of this assumption is indirect; other persons act approximately as I do under conditions when I know my self-awareness is present.

Scientific study of self-awareness is most difficult. The great physiologist
Sherrington (1953) said, "Though living is analyzable and describable by
natural science, that associate of living, thought, escapes and remains re-
fractory to natural science. In fact natural science repudiates it as something
outside its ken. A radical distinction has therefore arisen between life and
mind. The former is an affair of chemistry and physics; the latter escapes
chemistry and physics."

Sherrington was not a vitalist or a spiritualist; the statment about "thought,"
or self-awareness, or soul, should not be confused with Decartes' statement
that the soul "is of a nature that has no relation to the extent nor to the
dimensions or properties of the matter of which the body is made." Sherrington states the simple fact that self-awareness is something directly experienced by introspection, rather than observed or recorded by scientific instruments.

We accept our own introspective experience as evidence of self-awareness
in persons other than ourselves. When it comes to the problem of the existence of rudiments of self-awareness in animals other than man, evidence
by analogy becomes quite unreliable. Do dogs or horses or birds or insects
or protozoans have even the slightest traces of something like our own self-
awareness? Some scientists, like some lay persons, believe the answer is yes,
while others deny it. It is not necessary that we resolve this problem. What
is certain is that human self-awareness is at least quantitatively so different
from any traces of "mind" that can be present on the animal level that one

can regard it as a qualitatively different phenomenon. The possession of a mind makes the human species biologically unique.

The problem that inevitably presents itself to an evolutionist is, "How did self-awareness arise in human evolution?" It goes without saying that direct evidence bearing on this problem is absent. Yet indirect evidence is available, and some of it is quite convincing. Self-awareness is the basis of all human forms of social living. It serves to organize and integrate man's physical and mental capacities, by means of which he controls his environment. One considers most human beings responsible for their actions because one assumes that they are aware of themselves, and therefore of the consequences of their actions. Individuals who are insane are not responsible. Nor are animals, because one does not ascribe to them humanlike minds. Self-awareness has made mankind outstandingly successful as a biological species. It is the basis of the human version of adaptedness. We regard it as extremely probable that the genetic basis of self-awareness has been constructed by natural selection during the course of evolution. Yet only mankind, and no other species, has evolved this extremely powerful form of adaptedness. This means only that different organisms are adapted in different ways and is in no way contradictory to the postulate of natural selection.

Self-awareness is the cause of another unique human quality: death awareness. All animals die, but only man knows that his life is finite, and that some day he will die. Man is aware that he exists and aware of his personal future, which is inevitably death. An animal can observe death of conspecific individuals, but it cannot foresee its individual future far enough to discern death at the end. Man is aware of his self and therefore of his finitude; self-awareness could hardly exist in any meaningful form without death awareness. Conversely, it is hard to imagine awareness of the inevitability of death without self-awareness.

Death awareness leads man to perform actions that tend to leave traces in archeological and fossil records—namely, the ceremonial burial of the dead. The forms these ceremonies take in different human cultures are extremely diverse—interment, cremation, the exposure of cadavers to carrion-eating birds or beasts. What is found rarely, if ever, among human cultures is paying no attention to the dead, or the disposal of cadavers as if they were rubbish. In contrast, this is exactly what happens with nonhuman animals. They usually show no interest in dead individuals of their species: ants and bees throw them out of their nests with other rubbish; termites eat them like other food. Only some animals which have developed parental care of the young may continue it for some time after these young have accidentally died. In the case of human beings, ceremonial burial is extremely ancient. The Neanderthal race certainly made such burials; specially prepared graves have been found in which the bodies were laid, together with stone implements and adornments such as the skulls and horns of wild beasts. The discovery of patches of mixed pollen grains in such graves suggests that flowers were also placed with the corpse. The placement of objects in a grave suggests not only death awareness but also the possession of some belief concerning future life.

Whether or not a genetic basis of death awareness could have been a product of natural selection is an open question. It could have been, if it led parents to make provisions for the benefit of their progeny in anticipation of their own demise. This seems, however, to require a rather more advanced culture than is likely to have existed at the beginning stages of hominidization. Moreover, many insects provide for their progeny, whom they never see except as eggs which they deposit. There is certainly no reason to ascribe death awareness to these insects. Human death awareness may have been solely a by-product of self-awareness. The latter is certainly of high adaptive value, while this is questionable for the former.

The importance of death awareness for the cultural evolution of mankind has been enormous. Death seems a negation of any meaning of life. Many religious and philosophical systems have attempted to overcome this negation. The proposed solutions are numerous and diverse. Whether or not any of them is valid need not concern us here. What is important to an evolutionary biologist is that the roots of the religious quest lie deep in the genetic endowment of our species.

Ethics and Evolution

A poetic account of the decisive evolutionary step from animal to man is given in the Book of Genesis: "And the Lord God said, behold, the man is become as one of us, to know good and evil." Any human society from the most primitive to the most advanced has a system of ethics which it tries to impose on its individual members. Whether or not animals may have some rudiments of ethics is uncertain. For ourselves, we agree with Simpson (1969) that "it is nonsensical to speak of ethics in connection with any animal other than man. . . . There is really no point in discussing ethics, indeed one might say that the concept of ethics is meaningless, unless the following conditions exist: (a) There are alternative modes of action; (b) man is capable of judging the alternatives in ethical terms; and (c) he is free to choose what he judges to be ethically good."

The animal ancestors of the human species had no ethics. The questions inevitably arise, then, "What is the biological foundation of the capacity to have ethics?" and "How has this foundation been shaped during evolution?" The behavior of animals, no less than their anatomical and physiological characteristics, is a product of natural selection. Is it possible that human ethics and morals are of similar origin? Human individuals acquire their ethics from parents and other persons who bring them up. In other words, ethical codes are parts of cultural inheritance, not of biological heredity. It is not ethics themselves but the capacity to have ethics that has a biological foundation. The situation is analogous to what we found earlier with human symbolic language: the capacity to acquire a human language is genetic, but what language is acquired is a matter of culture. So it is with ethics: human genes are a precondition for the acquisition of ethics, but whatever ethical code is acquired is learned from other human beings and shaped by circumstances.

Is man by nature good or evil? This topic has been endlessly and inconclusively disputed by philosophers since time immemorial, and more recently by ethologists (students of animal behavior) as well. There have been theories of original sin, which claimed that human nature is intrinsically corrupt, and other theories which alleged that man is born good but is turned bad by vicious societies. Great ethologists such as Lorenz, Tinbergen, and their students believe that man has inherited from his animal ancestors aggressive drives, tendencies toward violence, and other such unpleasantries. An analysis of these claims lies beyond the framework of this book.

Our own position is, briefly, this: man is born neither good nor evil but with a capacity to become either; that is, to acquire whatever mixture of good and evil tendencies the circumstances of his personal biography induce him to have. To prevent misunderstanding, let us make clear that our view is not that man is born a tabula rasa, a blank slate. Some individuals in a given social environment learn more easily than others what is considered good or bad behavior. And these individual differences are, in all probability, under genetic control. This is not a belief in genetic determinism. In the first place, the manifestation of behavioral tendencies is influenced by the social environment; an individual who is likely to come into conflict with the law in one society may be considered a hero in another society. In the second place, in a given society, innate tendencies may be encouraged or overcome by training and education, which are not identical for different individuals. And in the third place, an individual may be able with some effort to control certain unwanted drives.

Even granting that our views are valid, there remain many facts that an evolutionary biologist must try to explain. Many kinds of behavior observed in some animals would be considered ethical or altruistic, and others unethical and egotistical, *if these behaviors were observed in human beings.* Thus, the behaviors of workers among ants, termites, and social bees and wasps strike us as models of altruism and unselfish devotion to the common good. In contrast, the behavior of dominant individuals toward subordinate ones of the same sex in herds of some monkeys and flocks of some birds seem to human observers to be selfish and egotistical. These varieties of animal behavior are clearly inborn and genetically conditioned; they are products of natural selection. Is it then not probable that some human behaviors, good and bad ones, are likewise established as fixed traits in human evolution? Thus, we come back once more to the problem of whether not only the capacity to acquire ethics but also some particular ethics are genetically conditioned. And if humans are not as dedicated to the interests of their societies as ants are to the interests of their colonies, does this mean that natural selection failed to do a good job in the human species?

Here we must introduce a distinction between what may be called family and group ethics. Parents are willing to suffer personal inconveniences and to make sacrifices on behalf of their children. Members of a family, as a rule, tend to help each other, often at a cost to themselves. Interestingly enough, children are, on the average, less anxious to help their parents than are parents to help their children. All these are family ethics. Probably in all human societies, parents who neglect their children are considered to act unethically; acting in the interests of one's family, on the contrary (but within prescribed limits), is usually regarded as "natural" and proper. With much less uniformity, in most societies individuals are also expected to help members of their society who are not their close relatives. At the limit, every human being will, hopefully, help any other human being in need of help.

Self-sacrifice on behalf of one's country, or of mankind, is admired as an act of heroism, a highly ethical form of behavior. These are group ethics.

Family and group ethics are notably different from the standpoint of biological evolution. Family ethics in man could be, and probably actually have been, promoted by natural selection. It is not accidental that forms of human behavior enjoined by family ethics have the most parallels among nonhuman animals. In man, as in many (though not in all) animal species, parents must be willing to take care of their progeny, or else the species will die out. To be sure, in some animals sex cells are released in ambient water, fertilization occurs outside the bodies of the parents, and the progeny are left to fend for themselves. But even in these animals the parents expend energy to produce an abundance of sex cells, and physiological mechanisms exist leading to the simultaneous release of male and female sex cells. There is no valid reason to expect natural selection to do anything to induce children to aid their parents, and such reciprocity of help is rare or absent among animals.

Haldane (1932), Wright (1949), and more recently Hamilton (1964) have analyzed the problem of "altruism," or family ethics, from an evolutionary point of view. Natural selection can reinforce altruistic, or self-sacrificial, behavior between genetically related individuals, provided only that this behavior helps to perpetuate and multiply genes present in the altruistic individual. Parents, especially aged and postreproductive ones, may well sacrifice themselves on behalf of their children, because the value to the population of a young individual who will become a parent is greater than that of an old one. Furthermore, since postreproductive parents can no longer introduce genes into the gene pool of the population directly, the best they can do is to better the lot of those individuals (their progeny) who carry their genes already. There are good genetic reasons why brothers and sisters should help each other; more tenuous ones for cousins to help each other; and no (genetic) reason why unrelated individuals should help one another. Yet human group ethics ordain mutual help among human beings regardless of genetic relationship.

Attempts have been made to explain the origin of group ethics by a special form of natural selection called group selection. Suppose that a species is broken up into numerous isolated colonies, tribes, or clans. Suppose further that some of the tribes contain some altruists willing to make sacrifices for the common good of the tribe and other tribes contain few or no such individuals. The former are likely to prosper and to become more numerous than the latter. The difficulty with this scheme is that group selection is a much slower and less efficient process than ordinary individual selection. When the unit of selection is the population or a tribe, it takes many more generations to have the genes spread to the rest of the species than when selection deals with individuals. The development of the marvelous "altruistic" behaviors among social insects is the only exception, and that exception confirms the rule. An ant or termite nest contains very few sexually active individuals; these individuals do all the reproduction in the colony, while the workers are sexually undeveloped and sterile. Now, it is particularly the workers that exhibit altruistic behavior on behalf of the colony. They can do so because the genes they carry as individuals are not transmitted anyway, while if their behavior benefits the sexually functional members of the colony the genes of the latter have an increased probability of being perpetuated. Furthermore, because the workers are generally the daughters of the queens, the genes involved are largely the same at any rate. Thus, the unit of natural selection in this case is the colony, not the individual worker.

In contrast to family ethics, group ethics are the fruit not of the genetic but of the cultural evolution of mankind (Campbell 1972; Dobzhansky 1973a). We believe that it is easiest to envisage the origin of group ethics as an extension of family ethics, eventually to all humankind. All human beings are "brothers and sisters" and ought to be treated as children of the same parents. The suffering of any human being anywhere ought to be relieved, as one would strive to relieve one's own or one's relative's suffering. Group ethics occasionally, but only occasionally, conflict with family ethics propagated by natural selection. For example, natural selection tends to favor maximum fertility of individuals. Today's overpopulated earth should induce people to accept the ethic of restriction of fertility by efficient forms of birth control. To some extent, group ethics should instill in persons an altruism like that which makes an ant worker put the interests of her colony above her own, but with these two cardinal differences: first, the altruism should be freely accepted rather than enforced; and second, it should favor the entire species rather than any subdivision thereof. The adaptedness of the human species would certainly be increased thereby.

Is the Human Species the Apex of Evolutionary Progress?

The uniqueness of man was recognized already by Aristotle, who wrote that man is the *"zoon politikon."* Evolutionary uniqueness is, of course, the property of most biological species. Evolution as a historical process is irreversible and nonrepeatable. A species arises only once. Allopolyploid species are perhaps exceptions: a doubling of chromosomes in a hybrid of two different species may occur in different places and at different times. Of course, when one speaks of the evolutionary uniqueness of mankind, one means more than the general uniqueness of all biological species. Some authors have claimed that mankind is the most advanced, most perfect, most progressive fruit of the evolutionary process. Even more: the purpose, the intent of all evolution was to produce man. The phylogenetic line leading to man is, in their eyes, the central and privileged one.

All such claims must be subjected to critical examination. We have tried to show in previous chapters that, according to the modern synthetic (often misnamed neo-Darwinian) theory, evolution is *not* programmed to produce man or any other particular species. It has no purpose, either, although the agencies that bring about evolution do frequently (but not always) yield an internal teleology which enables the organism and the species to survive and reproduce. A philosopher or theologian has a right to consider the evolutionary line of man a privileged one: it is the most meaningful one for us humans (Teilhard de Chardin, 1955, 1959). Yet a philosopher as well as a biologist must recognize that, until the emergence of superorganic culture, the fundamental causes that operated in human ancestry were the same as those of other evolutionary lines. Contrary to the assertions of some biologists (such as Grassé, 1973, 1977), biological evolution does not occur according to any preconceived plan and has no direction other than that provided by the perpetuation of the species. The high frequency of extinction disclosed by the paleontological record shows that even this one "direction" is by no means always maintained. This is precisely what one might expect, because the principal guiding agency of evolution—natural selection—has no prevision of the future.

Biologists, like other persons, have an intuitive certainty that some forms of life, such as vertebrate animals (including man), represent higher levels of organization than others, such as bacteria and blue-green algae. Nevertheless, no attempt to define what is evolutionary progress has thus far been successful (see Ayala 1974). One can legitimately speak of several forms of biological progress, provided that it is made clear which form is being considered. Increasing biomass, the bulk of matter incorporated into all the members of a species or of a group, is one kind of progress. The ancestors of our species were rare animals, restricted to a part of Africa. Human beings have since spread throughout the globe and have increased in number to more than four billion individuals.

Increasing complication of bodily structures and functions is another possible criterion of progress. Man's body is surely a highly complex machine, though not obviously more so than those of other mammals or of birds. The development of sense organs and nervous systems is still another criterion. Man's sense organs are good, but not the most perfect among living beings. For example, some birds of prey have keener vision, and many mammals have keener senses of smell. Man does have a remarkably large brain in proportion to body size. In brain power, rather than in body power, man unquestionably holds the foremost place in the biological world. Some biologists and some cyberneticists have emphasized the growth of the amount of information contained particularly in genetic materials. The application of this criterion may perhaps place man at the apex of the living world, but these measurements are, at the moment, somewhat ambiguous.

An interesting trend in evolution is toward increasing individuality, and the increasing importance of individuals. Among microorganisms, adaptation to drastic environmental changes, such as the introduction of antibiotics or transfer to altered nutrient media, may occur through the destruction of countless individuals. Populations adapted to new environments may be built up from one or very few individuals that happen to carry mutant genes making them viable in the new environments. The sacrifice of masses of individuals is minimized in more complex organisms, particularly in higher animals and, to a lesser extent, even in higher plants. Here the life of the individual is protected by a variety of homeostatic mechanisms that make natural selection occur more efficiently; that is, at a lesser cost of individual deaths. Culture and technology have maximized the value of an individual's life in the species *Homo sapiens*.

The human species is unique in the living world because of a complex of interdependent characteristics. Although some or perhaps all of these characteristics may be found as rudiments in nonhuman animals, as a complex they are the property of humankind alone. Here belong the ability to think in abstractions and symbols, symbolic languages, dependence on tools and tool making, and self-awareness and death awareness. These characteristics have given rise to a unique method of adaptation and control of the environment by means of learned rather than genetically transmitted culture.

One can hardly make the achievement of culture a valid criterion of biological evolutionary progress. This is simply because culture has been achieved in only a single species in the entire living world. And yet by this achievement human evolution has transcended, that is, it has gone beyond the limits of, biological evolution. Some biologists and philosophers would restrict the term "evolution" to biological evolution only. Others use the term in a more inclusive sense, comprising cosmic or inorganic, biological, and human evolutionary phases. Considering evolution in this broader sense, one may identify in it two major transcendences: that from inorganic to organic and that from organic to human evolution. Let it be clear that by "transcendence" we do not wish to imply any kind of philosophical transcendentalism or mysticism. Transcendence, going beyond the limits of the preceding evolutionary phase, means only an evolutionary event which introduces novel laws of nature. Biological laws, such as Mendel's laws, simply do not apply to inorganic nature; they do not, of course, contradict any law of physics or chemistry. Likewise, laws of human societies make no sense in the worlds of animals or plants; nonetheless, they do not abrogate any biological law.

CHAPTER 5

The Biological and Cultural Evolution of Man

From earliest human history, philosophers, scientists, and even ordinary persons have wondered about the origin of man. Ideas have never been wanting; many—diverse and contradictory—have been handed down in ancient documents. Seemingly immune to the infirmities of age, the ideological and scientific controversy over man's origin remains alive and well today. The Book of Genesis, an ancient account of ideas that was written down in Babylonia during the eighth century before the birth of Christ, tells us: "And the Lord God formed man of the dust of the ground, and breathed into his nostrils the breath of life; and man became a living soul."

In Genesis we have one of the two principal explanations of man's origin: a single act of creation performed at the will of a Creator. The Creation Research Society advocates this explanation even today in the courts of California, Arkansas, Louisiana, and several other states. As a signed stipulation, members of the Creation Research Society affirm that the biblical account of man's genesis is an absolute truth that must be accepted literally.

Opposed to the idea of a single act of creation is the evolutionary view, according to which man is the product not of a single creative act but of a long series of evolutionary steps. According to the evolutionary view, man was not *created* as we know him today but rather has *developed* into what he is today, and will continue developing into something still different in the future.

In the preceding chapters, we have completely accepted the evolutionary theory of the origin and subsequent development of the human species. In this and the following chapter, we shall focus on certain controversial aspects of evolution in greater detail. We do this because we feel that knowledge and attitudes based on knowledge are crucial for man's future. Cultural evolution, that which separates us most strongly from other animals, not only permits but also forces us to interfere with our own evolution. The debate over evolution, then, is not purely academic: *the outcome of that debate will determine man's future.*

Views on Human Origins: A Historical Survey

The evolution of concepts regarding man's origins will be presented here in sequence. Through the gradual replacement of faith by scientific knowledge as the latter has grown, one might guess that creationism would have been slowly but inexorably displaced by evolutionary ideas. The truth, however, is more complex than one might expect.

In prehistoric times, man had probably already formed ideas concerning his origin, his place in nature, and his future. The data on this matter are scanty. The interpretation of burial rites (recall that bouquets of flowers were placed in graves by Neanderthal man) and of certain prehistoric paintings suggest that human beings living thirty-five thousand years ago had both self-awareness and an awareness of death.

Written documents, some dating from the dawn of the historic epoch, reveal divergent views over man's origin and his place in nature. Correspondingly opposing views are found even now. In former times, the opinions and philosophies concerning the origin of man could hardly have been influenced by scientific facts—i.e., objective facts. Nor were the positions taken by various schools objective. The question of man's origins is not confronted by persons whose opinions are freely arrived at. Each spokesman is biased by constraints imposed by the time and place in which he works, and by his own social situation.

Intense scientific research on the origin of man has been undertaken only since the middle of the nineteenth century. Despite the mass of data that has been accumulated by paleontologists, archeologists, anatomists, embryologists, physiologists, biochemists, geneticists, psychologists, and linguists, no consensus regarding man's origin and his relationship to other organisms has been arrived at.

Under the pressure of scientific discoveries, most persons now accept (perhaps with some reticence) the idea that man, like all other organisms, is a product of evolution. Nevertheless, one still encounters positions that are diametrically opposed to the idea of man's evolution. Contrary to a commonly held belief, scientists are not totally objective. When the question concerns the origin of man, extrascientific constraints are particularly strong. Nor are these constraints entirely conscious. That there have always been creationists and evolutionists should come as no surprise. Among the former are both rigid fixists and others who admit to a limited evolution following separate acts of creation; among the latter a conflict can be observed between those who appeal to natural, knowable forces and others who appeal to unknown—even unknowable—forces in order to "explain" evolution. To better understand the differences among present-day positions, we may note that creationists and evolutionists have always been opposed.

One finds in the Vedas (religious and poetic texts of ancient India) and in the Upanishads (which date from the first millennium B.C.), for example, the idea of a development of a structured world from an initial chaos. The motive force created the world from substance. Later, Buddhism proposed a cosmogony which rested on an autonomous series of causes and effects: nature governs itself according to its own laws. Spiritual phenomena, in this view, are part of the unfolding of natural events. No single principle exists above or outside of nature. More than that: by studying nature, man can intervene in and control his own fate, thus participating in evolution.

At the opposite end from Buddhism is Brahmanism: spiritualist, fixist, and absolutely domineering. Brahmans considered themselves to be of divine origin, charged with the guardianship of other humans, who must blindly obey their orders. The Brahmans made themselves the masters of Indian society. They exploited a system of rigid castes, transforming it into the most disastrous and extreme form of fixity that ever existed. Their caste system protected their own dominant social position. Clearly, sociological concerns influence antievolutionist attitudes. In Brahmanism, Vishnu is the savior god, the active god, and the preserving god. "The law of transmigration immobilized and eternalized the inequalities of classes and races; it lessened the distance separating man and animals, but it did so by augmenting the distance that separates one man from another," according to the 1867 Larousse. Brahmanism is absolutely antievolutionist.

Taoism, originated by the Chinese philosopher Lao Tzu in the sixth century B.C., is based on a concept of continual evolution of the world. Lao

Tzu said of the Tao that it itself does not change but that it is the cause of all change, the source of the universe, of all that grows—the mother of everything.

Lieh Tzu, a successor of Lao Tzu, propounded in the fifth century B.C. a Taoism even more plainly evolutionary. In the initial chaos, energy, form, and substance were intermixed. Tao wrought development and uninterrupted transformation. The various forms of life have issued one from another. By the end of the third century B.C., however, the rigid and dogmatic concepts of Confucianism had become influential enough to challenge the evolutionary philosophy of Taoism.

In ancient Greece of the ninth century B.C. Homer claimed that the world and mankind had not been created but had developed. Democritus affirmed that nothing arises from naught, and that nothing can be destroyed. Everything comes from something that existed earlier; everything will become something different. Thus the celebrated expression "Panta rhei" ("all is flux") of Heraclitus of Ephesus—the best known of the Ionian naturalistic philosophers. The universe, in Heraclitus' view, represents the outcome of a spontaneous evolution. Anaximander, in the sixth century B.C., proposed an evolutionary origin for all organisms, including man; the whole world, according to his view, is continually differentiating.

The pre-Socratic philosophers, the cosmologists, undertook the task of explaining the origin of the universe. Not one of them, according to Büchner (1869), invoked other than physical, material causes in postulating a primeval material from which all else emerged. Not one of them acknowledged the dualism which was invented later: matter and spirit, body and soul. The early cosmologists were monists.

Other philosophers of ancient Greece, namely, the eleatics—whose school was founded by Parmenides and Xenophane during the fifth century B.C.— claimed, to the contrary, that nothing changes. An object, a being, is that which it is and never changes. If one entity were able to change into another, it would *be* and *not be* at the same time. That which *is* would be able, then, to change to something that it has not become—therefore, to that which *is not* at the moment. The eleatics viewed this as impossible. Development and destruction are impossible; both the transformation and the plurality of things are mere illusions. The only thing which exists and which remains eternally the same is immutable existence.

Plato, close in thinking to the eleatics, aspired to a stable world. What our senses make us perceive is only an illusion. True reality lies within general concepts, which are not corporeal. Individual human beings or horses, taken one at a time, have different appearances. Their existence is of short duration. The concept "human being" or "horse," however, neither changes nor perishes. For that reson, concepts are the true reality. For Plato, the *eidos* represented the sole reality, pure and immutable. The philosophy of present-day typologists, who do not accept the real flux of evolution (and they do not, as can be seen if one scrutinizes their work carefully) can be connected by intermediate stages to the objective idealism of Plato.

For Plato, as for the Hindus and Christians, the perfection—the ideal— is given from the beginning, and is not the product of evolution.

Without entering into all the richness and diversity of the Greek philosophies, it is interesting to emphasize the two principal, diametrically opposed philosophical positions: the belief that man and the entire cosmos have been created as they are and remain immutable and the belief that man, other living beïngs, and the universe itself have changed during the course of time. All is flux: *"panta rhei."*

On another essential point we again find among the Greek philosophies two opposing tendencies. For the one (notably for Plato) there was a duality of ideas: that of the spirit and that of material. For the Greek philosophers who preceded Plato, spirit and soul were material, not to be distinguished from physical matter. That opposition between materialists and idealists, between dualists and monists, exists today just as it did among the ancient Greeks—and earlier.

Just prior to the Common Era, the Latin poet Lucretius presented his ideas concerning the progressive development of the human species and of civilization. Our first ancestors lived like animals, according to Lucretius. Little by little, they learned to build huts, to dress themselves in skins, to use fire, and to develop language, art, and inventions.

Scientific observations and materially established facts had little, if any, influence on the philosophers of whom we have spoken. Whether one held evolutionist or creationist views depended above all on one's origin, one's temperament, or one's sociological constraints.

Even though the ancient Greeks knew of and even described stone axes and other weapons (which they called ceraunia), they did not regard them as man-made products reflecting man's cultural evolution; the Greeks regarded such objects as the products of lightning bolts. Even during the renaissance of agriculture, these artifacts were regarded as freaks of nature.

Michel Mercati is probably the first to have proclaimed, near the end of the sixteenth century, the true nature of ceraunia—followed by Boetius de Boot (1636), Aldrovandus (1648), Hassus (1714), Jussieu (1723), and others. Boule (1946, 1957) said, "Nevertheless, although the earlier thinkers may have known that historic civilizations had been preceded by uncultured or savage periods, they had no notion of the enormous antiquity of these primitive times. One had first to accommodate theories to the requirements of biblical chronology."

Buffon (1749) is probably the first who, in opposition to church dogma, had a sense of the immense duration of time. He claimed that the immense quantity of marine fossils that are found in so many localities proves that they could not have been transported by a single flood.

Concerning the history of man, Buffon (1749) claimed that there was originally only one species of man, which, having multiplied and spread over the entire earth's surface, underwent different changes under the influence of local climates. The varieties of the human species, according to Buffon, are perpetuated from generation to generation, just as the deformities of fathers and mothers are often passed on to their children. Buffon concluded his account by saying that, through the gradual disappearance of certain of them, present-day varieties will come to differ even more than they do.

The deputies and trustees of the Faculty of Theology suggested to Buffon that his Natural History included principles and maxims that did not conform to religious ones. They censured his work, and on March 12, 1751, Buffon prudently retracted his earlier remarks in the following terms: "I declare that I had no intention of contradicting the text of the Scriptures; that I believe very firmly all that is reported there regarding creation as to both temporal order and the circumstances of events; and that I abandon that which, in my book, concerns the formation of Earth and, generally, all that which seems to be contrary to the account of Moses." We may recall that creationists in many states have exerted ever stronger pressure to obtain for biology teachers the same censure as that applied to Buffon by the theologians of the Sorbonne in 1750.

Only in the nineteenth century were fossilized skeletons of humans and prehumans first found. One might think that with this material proof, the matter of man's evolution would at last have been settled. Well, it has not been! With respect to man's origins, objectivity is less than perfect and preconceived ideas reign.

In his *Philosophie Zoologique* (1809), Lamarck had already postulated without reservation that man was a product of evolution, just like all other organisms. But his colleague at the National Museum of Natural History, the baron Cuvier—a man who was avid for power, and who denigrated his colleagues—insulted Lamarck in the funeral eulogy he delivered at the Academy of Sciences by propounding on that occasion a thesis diametrically opposed to that of the deceased. Cuvier was a spiteful dictator in biology, and his influence extended, through his students and his students' students, even beyond the nineteenth century.

Cuvier was absolutely convinced of the constancy of species. That he was is all the more astonishing because his paleontological research (whose results are summarized in five volumes entitled "Research on Fossil Bones," published in 1812) could have made him the first evolutionist; his extensive observations could have led him to a transformist theory. The study of the succession of fossil species during the course of geological time should have prepared him in a particularly adequate manner for an evolutionary concept. Here are his words, taken from the introduction to his paleontological work (Cuvier, 1978): "Would it not also be glorious for man to burst the limits of time, and, by a few observations, to ascertain the history of the world, and the series of events which preceded the birth of the human race? . . . And why should not natural history also have one day its Newton?"

In speaking of the facts he had uncovered, Cuvier displayed his vanity by claiming that "many of them are decisive; and I hope that the rigorous methods which I have adopted for the purpose of establishing them, will make them be considered as points so determinately fixed as to admit of no departure from them" (1978). One can only regret that the impressive mass of excellent descriptions of fossils which Cuvier compared with living species led him to the astonishing theory of the succession of fauna and of species in the course of geologic periods by means of destructive cataclysms, each followed by a new creation.

Cuvier claimed that man as we now know him had been regenerated after a worldwide flood that occurred some 6,000 years ago. He considered the Brahman tablets, which attest to the great antiquity of the peoples of India, to be inventions of their priests. "One can understand," he said, "what can become of history in such hands as theirs . . . their Vedas have been drafted . . . at the beginning of the present age . . . their astronomical tables have been calculated after the event, and badly calculated. . . ."

Traces of the flood—the one, true flood described by Moses—he found in Indian, Chinese, and Egyptian documents and even in the crude hieroglyphics of the Americas. In order to make data and dates conform, Cuvier interpreted according to his fancy; he rewrote texts and he accused "the sacerdotal races of India of having arrogated to themselves sovereignty over the mass of the people," an act that permitted them to rewrite history according to their wish to dominate. He also added that "the most degraded of the human races, that of the Negroes, has the most nearly brute-like form . . . and has preserved neither records nor tradition. They have been unable to contribute to our search."

In a discourse of some sixty pages Cuvier selected, eliminated, and rewrote historical facts of all geographic regions in order to conclude that "the surface of our globe has been the victim of a large and sudden revolution whose occurrence could not be more remote than five or six thousand years. . . ."

Cuvier lived during the same era as, worked in the same environment as, and was perfectly acquainted with the work of Lamarck and of Geoffroy Saint-Hilaire. He thus developed his ideas in direct opposition to those of his two evolutionist colleagues. Cuvier, consequently, represents a particularly convincing example of the influence of preconceived ideas and personal circumstance on the interpretation of data regarding the origin of man. Cuvier, the son of a Protestant officer, having known difficult beginnings, became a member of the Academy of Sciences, Professor of the College of France, Chancellor of the University under Napoleon, State Counselor under Louis XVIII, Baron, and Grand Officer of the Legion of Honor, and was granted a peerage under Louis Philippe.

Lamarck was descended from a noble family, joined the military, and then became a botanist and zoologist. He also became a member of the Academy of Sciences, but otherwise confined himself to the chair of Zoology at the National Museum of Natural History. Lamarck was one of those who are passionate with respect to their research. He took no part in intrigues, in closet activities, or in the submissions and compromises which lead to the false grandeur of public acclaim. Without a doubt, the price he paid for his independence permitted him to become the father of transformism. Cuvier was not himself characterized by a breadth of spirit that would have granted him the courage and vision of an entirely new concept.

Cuvier, and subsequently many others, reproached Lamarck for not having founded his theory on observational facts, and for having presented only wild speculations. Such accusations are not true. Lamarck was led to his theory of transformation by an extraordinary wealth of personal research on the important invertebrate collections of the Museum of Paris. Cuvier, for his part, possessed a profound personal knowledge of fossils. Both men possessed substantial facts on which to take a position in the debate on the origin of species.

The direct proofs (embodied in fossil remains) concerning the origins of man were still lacking at that time. Hypotheses could be only speculative. The analogy with other organisms nevertheless provided both with excellent arguments. We find that in the same era and with the same excellent scientific training, Cuvier arrived at a totally erroneous speculation, whereas Lamarck had the correct view.

In the second half of the nineteenth century, and above all after the shock created by Darwin's "Origin of Species," research on evolution and on its mechanisms was undertaken. The theory of evolution of lower organisms reinforced the evolutionary position in discussions of man's origins. The discovery of skeletons of real human precursors, of prehistoric utensils and paintings, supported more directly the evolutionary theory.

Rare are those who still hold openly to the unchanging and creationist point of view. We can, for instance, cite the example of the French Jesuit Descoqs (1944) who strove to demonstrate that transformism is in deep trouble. He did not fail to cite Lemoine, who claimed in the 1937 Encyclopédie Française that the theory of evolution is impossible and that, despite appearances, people no longer believe it. He also mentions Bounoure (1939), who said that merely noting the constancy of living species restores to honor the modern biology that has escaped from a tyrannic and out-of-date evolutionism. Descoqs also refers to his colleague Teilhard de Chardin, claiming that the continuity between diverse species which Teilhard regards as an a priori condition for intelligibility is totally without foundation.

Descoqs recalls a canon of the Commission of the Faith, "If anyone denies that the entire human race is descended from a unique first ancestor only, Adam, he shall be accursed," and adds that no one has the right to lay a hand on that thousand-year-old belief. Paleontology has not proven and will never be able to prove the finding of the first representatives of the human race. Only revelation permits us to infer with certainty, concluded Descoqs.

Among biologists, anthropologists, and paleontologists, there now exists a consensus concerning the basis of the origin of man by evolution. Nevertheless, divergences are considerable when the question involves an evaluation of the differences between man and other animals, their degree and significance, the mechanisms of evolution, and the continuity of evolution. By oversimplifying matters, one can place different opinions into three groups:

1) Some emphasize the enormous differences between man and other animals, also emphasizing our inability to understand or explain those evolutionary mechanisms which have made man a cultural animal.

2) In direct opposition to the above, others minimize the differences between the other animals and man, who, they claim, is nothing but a naked ape.

3) A third group (to which we belong) recognizes the important differences between *Homo sapiens* of today and all other animals, but claims that we are capable of understanding the mechanisms of evolution of all organisms—including man.

Even among these groups, preconceived ideas outnumber scientific facts in determining the positions adopted by various authors.

Darwin did not speak explicitly of the origin of mankind in his *Origin of Species*. Through several testimonials, however, we know that there was no doubt on Darwin's part that man is a product of evolution.

In one conference, Haeckel (1882) testified with reference to his meeting with Darwin in October 1866 at Darwin's home in Down. Haeckel had previously shown him the proofs of "General Morphology." In that work, Haeckel admitted that he had tried to found a science of organic forms on the basis of Darwin's theory. Prudence, says Haeckel, caused Darwin not to mention the anthropological consequences of his theory when writing the *Origin of Species*. In 1866, however, Darwin said that the application of his theory of evolution to man was necessary. Evidently Darwin had conceived of the evolutionary origin of man well before 1871. He had, in fact, claimed earlier that the *Origin of Species* would throw light on man's origins. After

Darwin, writers either dwelt on the resemblances between man and higher animals (notably the apes) or emphasized the importance of the differences.

Lyell (1863) spoke of the enormous gap which separates man from brutes. No one, he claimed, could imagine an animal developing a sense of happiness through faith in a life eternal. Only man has that faith. Lyell concluded that, far from having a materialistic tendency, the introduction into the earth of life, sensation, instinct, the intelligence of the higher mammals, and finally the improvable reason of man himself suggests an ever-increasing dominion of mind over matter.

Lyell, who had a strong influence on Darwin (notably through his concept of gradualism in geology), favored a unique position of man; he insisted on a hiatus between man and all other animals, but imagined a progressive evolution extending from the first sensory perceptions of primitive animals up to the development of human reason.

In comparison, we can examine the position expressed by T. H. Huxley in three essays written in 1863, still eight years before Darwin published his *Descent of Man*. In the second essay he posed the problem directly: "The question of questions for mankind—the problem which underlies all others, and is more deeply interesting than any other—is the ascertainment of the place which Man occupies in nature and of his relations to the universe of things."

With a characteristic candor and freshness of spirit, he added: "Most of us, shrinking from the difficulties and dangers which beset the seeker after original answers to these riddles, are contented to ignore them altogether, or to smother the investigating spirit under the featherbed of respected and respectable tradition."

Huxley affirmed that the structural differences between man and gorilla are large and significant: there is no intermediate link between *Homo* and *Troglodytes*. Similarly, there are no intermediate links between the gorilla and the orangutan. The same physical causes and biological processes that have created the different animal species have created man.

Huxley (1863) posed the problem clearly: "Our reverence for the nobility of mankind will not be lessened by the knowledge, that Man is, in substance and in structure, one with the brutes; for, he alone possesses the marvellous endowment of intelligible and rational speech, whereby, in the secular period of his existence, he has slowly accumulated and organized the experience which is almost wholly lost with the cessation of every individual life in other animals. . . ."

The zoologist Carl Vogt (1863) may also be cited. He gave a series of public lectures on man, in which he said, "No one denies the considerable distance that separates man from apes. . . . Nothing, moreover, should surprise us less than to find some marked differences between the anthropomorphic apes and man, because where would we find a line of demarcation if these differences did not exist? Still, it seems to us that these differences, as many as are found, in no way exclude a common type attached to a common root."

To study the relations between animal species and to discuss the position of man, he added, are not equivalent: "All the pride of human nature revolts at the thought that the king of creation can be treated as an object of nature. . . . The studies of man constitute a subject of research in which one wants to prescribe and impose on science the result to which it should necessarily arrive."

Vogt (1863) conceded the derivation of human species from apes while affirming "the large differences which separate them today, and which will continue growing as a result of civilization and progressive development of the human form. . . . By the incessant work of his brain, man has raised himself little by little and has left his barbarous state. . . ." At his ideological adversaries Vogt launched the following barb: ". . . it's better to be a perfected ape than a degenerate Adam."

In *The Descent of Man,* Darwin (1871) claimed that there is no fundamental difference between man and the higher animals in their mental faculties. He insisted on the importance of individual differences with respect to mental faculties, for such differences lead to natural selection.

Darwin thought that the mental faculties of man were developed gradually, by small evolutionary steps; therefore, the differences between present-day man and the most highly evolved animals have grown immensely during the course of long prehistoric periods: "The difference in mind between men and higher animals, great as it is, is certainly one of degree and not of kind."

Lower animals feel pleasure and pain, happiness and sadness, as does man himself. The majority of complex emotions are found in both higher animals and man. Dogs can exhibit jealousy. Animals not only are capable of love but desire to be loved, as well. Dogs show shame, happiness, modesty, and fear. Apes dislike being mocked.

Imagination is one of the most exalted perogatives of man. Because dogs, cats, and other animals have dreams, however, they must also have some capacity for imagination. To some extent, animals also reason.

Darwin took issue with those who insisted on separating man and inferior animals on the basis of mental faculties. It is claimed, said Darwin, that man alone is capable of a progressive improvement, of making tools and fire, of domesticating other animals, of having the property of using language; that no other animal possesses self-awareness or has the capacity of abstract thought or of general ideas; that only man has a sense of beauty or is capable of fantasies or has feelings of gratitude or mystery or believes in God and has a conscience.

Darwin responded that animals are capable of improvement and of learning during the course of their lives through individual experience even with respect to moral qualities. Animals, notably the chimpanzees, utilize objects which can serve as tools, but they do not manufacture them. Darwin admitted that that was a nonnegligible difference but asked whether man, in early times, might not have behaved similarly. Could he not, at first, have used some stone splinters that were obtained accidentally and then, little by little, developed the means for fabricating tools? We know that the improvement of tools generally occurs very slowly.

Regarding language, Darwin admitted that this faculty is one of the chief distinctions between man and the lower animals. Articulated language is found only in man, but like lower animals, he also uses cries and movements of facial muscles to express certain feelings. The ability to articulate is not that which distinguishes man from other animals, since the parrot is capable of speaking. Rather, it is the highly developed capacity in man to connect certain sounds to definite ideas, a capacity that depends on the development of the mental powers, that sets him apart. Speech is certainly not a question of instinct, for all languages must be learned. However, the tendency to *wish* to speak is instinctive, as can be verified among infants. Nevertheless, no language has been invented deliberately and in a single stroke. Philologists agree that all languages have been developed by small steps. Darwin recalls that even among birds a certain apprenticeship is needed for clear expression, and he mentions the local dialects that are found among birds and correspond to the distinct dialects of man. Darwin concluded: "I cannot doubt that language owes its origin to the imitation and modification of various natural sounds, the voices of other animals, and man's own instinctive cries, aided by signs and gestures. . . . May not some unusually wise ape-like animal have imitated the growl of a beast of prey, and thus told his fellow-monkeys the nature of the expected danger? This would have been a first step in the formation of a language." Man's articulated language is not an insurmountable objection against the thesis according to which man has developed from inferior forms.

Darwin declined to discuss abstract throught, individuality, and self-awareness, because these concepts are not adequately defined. No one supposes that an inferior animal reflects on its origin or its future—on death and life, for example. As Darwin asked, however, "But how can we feel sure that an old dog with an excellent memory and some power of imagination, as shewn by his dreams, never reflects on his past pleasures or pains in the chase? And this would be a form of self-consciousness."

Regarding faith in the existence of God and religions, Darwin claims, "There is no evidence that man was aboriginally endowed with the ennobling belief in the existence of an Omnipotent God." In fact, there are numerous data showing that many savage tribes possess neither concepts nor words expressing the idea of the existence of one or more gods.

The rational explanation of evolution by natural selection has caused heated debates, especially when the theory is applied to human beings. A familiar argument against Darwinian evolution that has been heard for more than 100 years concerns our ignorance with respect to the mind and conscience.

Du Bois-Reymond (1892) declared not only that the human conscience cannot be explained now but also that it never will be explained, that it lies outside the realm of science. He declared that science is and will always be incapable of understanding mind and conscience in terms of material conditions. He set limits to research beyond which scientists could (and should) not go: When a question involves the mystery of the essence of matter and energy, and of the capacity to think, science should once and for all adopt the slogan "We shall remain ignorant!" ["*Ignorabimus!*"]

More recently, Portmann (1947) reiterated that message in fixing insurmountable limits to biological research. Even the amoeba, he claimed, has not yielded to us the "mystery of life." We know a great deal, but we should occupy ourselves more with the limits of knowledge, with frontiers that logic should not exceed.

For Gagnebin (1943), "there is no more doubt: man has emerged by descending from a branch of anthropomorphic apes. . . . What were the causes of that evolution? We know exactly nothing. The obscurity remains complete. Similarly, regarding the factors and the processes of evolution, our ignorance is almost total. . . . We should be aware of the great mysteries of life and its history."

Again, Grassé (1973, 1977) declared that hominization seems to have been an evolutionary process that was not extraordinary in its unfolding but which has taken on an immense importance by its consequences. The modifications of the brain, he continues, whether qualitative or quantitative, have been the grand theme of the evolution of the hominids. They have conferred on us conscience and reason, from which is born free will.

Grassé does not doubt the evolutionary origin of man. He was even rebuked by Descoqs (1944) because he dared to say in his course at the Sorbonne in 1943: "To tell the truth, there is no more doubt concerning evolution, an idea without which Nature would remain entirely unintelligible . . . but whose mechanism remains hypothetical. None of the proposed mechanisms, and these have been numerous, give an exact account of the facts and observations. The puzzle of an organism's adaptation to its environment, to circumstances, remains inscribed on the lintel of the evolutionist temple."

In his recent book, Grassé (1973, 1977) attempts to demonstrate the inability of current doctrines to explain evolutionary phenomena. He believes, for example, that evolution has slowed its course, that it resulted previously from other causes and other mechanisms. Cosmopolitan man lives under all climates. The genotypes are quite different. Nevertheless, one sees no evolution. The two European races of the late Paleolithic—Cro-Magnon and Chancelade—closely resembled today's Europeans. "Thus, since the Aurignacean, persons who have inhabited and do inhabit Western Europe have at the most changed some minute details of their anatomy. . . . They have undergone no variation of any amplitude. . . . Mutations differentiate individuals . . . but in the real world the human species proves to be anatomically and physiologically stable." Today's Mayans resemble the portraits of their pre-Columbian ancestors. Chinese remain Chinese.

Now, relict species mutate as much as others. Despite microvariation on

the one hand and species stability on the other (extending in certain instances for hundreds of millions of years), must we conclude that microevolution has had no part in the true processes of evolution? Mutations,
Grassé's complaint goes, as numerous as they are, do not instigate an evolution of any importance. Of course they don't, because there is no correlation between mutations and evolution. *Mutation is not evolution.*

Evolution and the Domain of Science

The problems posed by evolution greatly surpass the analytic ability of
modern science. Our understanding fails notably in the knowledge of factors
that orient evolution according to certain "plans." Grassé doubts whether
we shall understand the causes of evolutionary lines, of the finality of structures, and of vital cycles. He concludes, "It is possible that in that domain,
biology, powerless, must yield to metaphysics."

While admitting the fact of evolution, Grassé proclaims that we cannot
elucidate its basic causes through either biology or biochemistry—neither
the essence of its plan, nor its motivating force, nor the linkages governing
its coordination.

We do not share Grassé's views. In the past as today, there have been
creationists and evolutionists; there have also been optimistic and pessimistic
rationalists—those who think that we shall understand human evolution better and better, and those who deny that we shall, or can. We think that
the causes and the mechanisms of all organic evolution, including that of
man, are accessible to scientific research.

Science itself has evolved and will continue to evolve as long as man
exists. Knowledge grows and accumulates, but not through simple addition.
The greatest proportion of established scientific "truths" are so only temporarily and conditionally. That is particularly true in complex scientific fields
such as that of organic evolution, especially that of human evolution. Observations still unmade will not simply be attached like wads of gum to those
already available. On the contrary, new observations will expose existing
knowledge to a new light and will call for a reinterpretation of today's knowledge.

Our knowledge of human evolution is obviously limited. But an attempt
to set limits to research, limits that cannot and should not be exceeded,

seems absurd. Such an attempt is as inadmissible as the claim that everything about evolution is already known.

Organic evolution presents an essential peculiarity, not found in other sciences, which sets limits of fact but not of principle. Each evolutionary step is a historical event that has never occurred before and will never be repeated. Experiments cannot be repeated in the laboratory until the underlying mechanisms are understood. Even for a perfectly reproducible experiment in chemistry, however, the degree of understanding was not the same in the time of alchemists or even of Lavoisier as it is today. The future, in turn, will improve our interpretation of chemical reactions, even those which we think we already understand.

The study of evolutionary successions is much more difficult than the study of a chemical reaction, which can be endlessly reproduced in the laboratory in seemingly identical fashion. Because each event is unique and because large numbers of factors intervene, extremely complex interactions arise. Is that sufficient reason, however, for fixing limits to research, for appealing to unknown and unknowable forces, for declaring that the determining factors of evolution were once entirely different and that the causes and outcomes of evolution escape us, and for postulating a predetermined plan and, consequently, a planner—two totally impenetrable proposals? We respond to those who reproach us for studying the great problems of evolution with seemingly naive faith in science: On what reason, on what observation, on what deduction does one base a gratuitous affirmation of unknown factors, of a predetermined plan, or of particular supraspecific evolutionary forces?

Research on the evolution of man is still young; many questions are still open. The best method of increasing our knowledge consists of noting observable evolutionary steps; using historical documents, fossils, tools, and cave paintings; and, for human evolution, using a reasoning-by-analogy method based on the known evolution of bacteria and *Drosophila*. Obviously, we can directly observe only infinitely small evolutionary steps—*micro*evolutionary steps—because *macro*evolution requires millions of years.

The situation for the study of evolution does not differ from that in the case of sedimentary rocks that give rise to mountains, or of the erosions that are subsequently responsible for their removal, or of the abrasion of granite rocks by glaciers. All scientists admit the possibility of explaining large geological phenomena by means of observations made on small modifications occurring before our eyes. We emphasize once more, however, that biological evolution is much more complex than geological evolution. Even so, there is no reason to overestimate difficulties and then suggest that limits be set to research, or to refuse to recognize interpretations based on actual observations, replacing them with mysterious forces such as undefined "systemic mutations" and other imaginary evolutionary processes. Even Lamarck recognized that genera, orders, families, and classes, categories that many claim need exotic explanations, are products not of nature but rather of systematists; higher taxonomic categories are merely categories of convenience which are useful, even necessary, for the classification of living things.

That man is a product of evolution in no sense implies that he does not differ from other animals. Among diploid, sexually reproducing organisms, each genotype, each individual, and each species is unique and temporary. Each represents a unique production: it never existed before and it will never exist again. Living material, whose individual forms exist only temporarily, is continually entangled in a process of transformation and therefore in the creation of evolutionary novelties.

In one sense, even the concept of species is a typological fiction. The species is not an *eidos*, definite and eternal, that is more or less realized in each individual. A species is a collection of individuals each of which differs from all others; individuals are *not* representatives of an immutable entity called "species." All species continually evolve. Even relict species that, in some instances, have remained seemingly unchanged for 100 million years or more have certain morphological traits that differ from those of their fossilized ancestors.

Such considerations are important for the interpretation of human evolution. Man is also not an *eidos* created once and thereafter destined to persist eternally exactly as it was, and is. Man does not exist apart from those individual beings known as persons. Furthermore, no two persons are alike: children are not identical to their parents, and even siblings differ from one another. The human species, like every other biological species, is a collection of dissimilar mortal individuals who, throughout past millennia, have succeeded in passing their slowly evolving genetic endowment from one generation to the next.

By no means do we minimize the numerous and very important differences that distinguish today's human from today's chimpanzee. These differences are the product of two long, independent courses of evolution. Man certainly was not created at the outset just as he is now. Neither language, nor abstract thought, nor self-awareness, nor awareness of death, nor the sense of being part of the totality of humankind, nor a sense of history, nor religious feelings, nor the capacity to make and use tools, nor the capacity to create works of art have always existed among "men." In that sense, the term "man" which does not allow for his evolving through the course of time is a typological fiction. In placing modern man among other organisms, one's concepts must be shaped in the light of evolution.

Another aspect of the fundamental (even sterilizing) error of the typological concept of the term "man" must be mentioned. For each characteristic of modern man mentioned above, there exists within any one population enormous variation between individuals exhibiting extreme values. These intrapopulation differences are greater than those between the averages of different peoples—averages which are, by definition, abstractions. This point will reappear in discussing human equality and genetic diversity.

A comparison of the capacities and characteristics of a Leonardo da Vinci, an Albert Einstein, or an Anatole France with those of a chimpanzee reveals the impressive differences separating a cultivated product of man's cultural evolution and an ape whose ancestors, for biological reasons, did not undergo cultural evolution. That comparison, however, does not prove that man by nature differs fundamentally from animals. Because these differences are now large, one need not conclude that they arose suddenly, in toto, during the moment when man was created. Never was there a sudden creation by a Supreme Being of something which is now called "man," a something which is fundamentally different from all other organisms.

Our ancestors, at some early time, were in no respect anything other than apes. The comparison between chimpanzees and present-day men is that between species which have long been separated. The kinship of the two, nevertheless, is much closer than that of many sibling species, or, for example, of different species of warblers.

Those who would establish a gap, or erect a barrier, between man and brutes (as they called animals even as late as the nineteenth century) forget the numerous fossilized transitional forms and species which are now known. Chimpanzees and other anthropoid apes form an evolutionary branch that has been differentiated from that of the hominids for about twenty-five million years. The only surviving species of the hominid line is *Homo sapiens*. However, there is in our branch, starting with the dryopithecids (a common ancestor with the great apes), a series of higher apes (including *Ramapithecus* of India and *Dryopithecus* of Africa) that will without doubt be extended and filled in during years to come. Starting from *Australopithecus* there is a differentiation toward present-day human beings that passes (using bold strokes) through the stages of *Homo habilis* (a tool maker), *Homo erectus* (a pithecanthropus), and that primitive *Homo sapiens*, Cro-Magnon, who can be considered our direct ancestor.

Today's man with his culture and his civilization is indisputably different from the nearest species among the anthropoid apes. That difference has been formed gradually, however, during the course of a long evolution, a gradual origin that is easily explained by mechanisms that are consistent with the biological theory of evolution.

The comparison between modern man and the chimpanzee reveals the creative power of evolution and the inherent potentiality of the path that was taken by man's ancestors; in no sense does the comparison prove the existence of an unbridgeable moat between man and other animals. Unless, that is, the evolutionary history of man is deliberately ignored.

Similarities and Differences Between Biological and Cultural Evolution

To incorporate the process of humanization into the general evolution of organisms, one must appreciate cultural evolution in its own right and grasp its relation to biological evolution.

The mechanisms of biological evolution do not allow rapid transformations—say, on the scale of a human life. Only ideas and concepts have evolved considerably during the last forty years.

The basic elements of biological evolution are the modification of genes through mutations—that is, changes in the sequences of DNA; the creation of new genes by the elongation (duplication) of DNA molecules; and the creation of new genotypes by segregation and recombination.

Such modification of genetic information does not in itself guarantee evolution. The latter requires changes in the frequencies of genotypes through the course of generations. Chance events, described by Sewall Wright under the name "genetic drift," also play a role. The principal evolutionary factor, and the only one that ensures a feedback relationship—an adaptive relationship—between environmental conditions and the capacities of the genotypes, is natural selection. Natural selection permits an orderly evolution and a harmonious coordination between coexisting organisms and between organisms and their environment. New mutations, considered to be rare events, are immediately tried by natural selection. New favorable mutations are probably extremely rare. Furthermore, a large number of generations are required for a new favorable allele to totally replace an ancient one of inferior selective value.

The study of the genetic structure of natural populations has greatly changed our ideas concerning the mechanisms of genetic evolution. We now know that populations are extremely heterogeneous. They show a high level of polymorphism, and their individual members are heterozygous for a sizable proportion of all gene loci. Because of this polygenotypism, modifications of genotypic frequencies can take place much more rapidly than was imagined even fifty years ago. Nevertheless, biological evolution, involving alterations in genotypic frequencies and the creation of new genotypes, remains a slow process.

The seasonal fluctuations of the frequencies of different genotypes observed by Timofeeff-Ressovsky (1940) among the coccinellids (ladybird beetles) or that described by Dobzhansky (1943) for *Drosophila pseudoobscura* are exceptionally rapid, requiring only several generations. The adaptation of Lepidoptera by a mimetic coloration (Ford 1971) to atmospheric pollution, known as industrial melanism, has required one or two hundred generations. The more profound anatomical or physiological modifications, however, are much slower in *Drosophila*, in man, and in all other organisms. From the point of view of biological evolution, the morphological and anatomical differences between modern man and Cro-Magnon man (or even earlier hominids) are not very prominent.

The evolution of the hereditarily fixed aspects of behavior, or of genetically determined "instincts," seems to be equally slow. One cannot pass judgment directly, however, because behaviors do not leave fossilized traces. Indirectly, the slowness of the evolution of genetically determined behavior can be verified by the comparison of nuptial displays and vocalizations in related species of ducks, of *Drosophila*, of crickets, or of frogs. The specific differences are slight; therefore, one can conclude that the evolution of behavior is slow.

For an adequate comprehension of hominization, the relation between biological evolution and cultural evolution must be discussed. Some persons have tried to make a distinction between the two, by which biological evolution would be the result of new mutations or changes in the frequencies of alleles under selective pressures and cultural evolution would concern exclusively and independently that which has been learned by each individual during the course of his life. This simplistic and erroneous scheme has also been proposed for the separation of innate and acquired behaviors.

No behavior is totally fixed by a genetic determinism; nor are acquired behaviors ever totally free of the genotype. *The capacity to learn depends on the genotype.* All behavior is the resultant of genotypic and environmental influences, learning experiences, and even freely made, arbitrary decisions.

The role of these interacting elements is not the same for different types of behaviors, nor for different individuals of one species; furthermore, it differs considerably between different groups of animals according to their evolutionary positions.

Genetically determined instinctive behavior is important for the female digger wasp (*Ammophila*), who prepares a hole, finds a caterpillar, immobilizes it by stinging the proper nerve ganglia, lays eggs on it, buries it in the hole, and refills the hole. For the crow, dog, or chimpanzee, genetic determinism plays a considerable role in behavior. The capacity to weigh circumstances, to react by voluntary choices, and to enrich the repertoire of behavioral reactions through learning is important among higher animals. The same is true for man. Human behavior is not totally and entirely a product of free will. The part played by genetic determinism in our behavior may, even so, be larger than many think.

A progressive evolution, extending from primitive animals to man, in the part played by learning in behavior, a part which grows hand in hand with the growing complexity of the nervous system, can be demonstrated. The enlarged capacity for individual learning, much greater in man than in other animals, was a necessary but not a sufficient condition for man's extraordinary cultural evolution. In animals (and without doubt this was so in early hominids as well), each individual must commence its apprenticeship from point zero; the young profit by imitating those things that adults have learned.

What gives man an extraordinary supremacy, what has permitted him to evolve at an unprecedented speed, is the transmission of learned experiences by means of an articulate and reflective language. We shall return to this shortly. The unique manner of transmitting learned knowledge was further enhanced by the invention of writing. Through writing, man has been able to conserve knowledge and to spread it through space and time. From the invention of writing, from the possibility of accumulating all individual experiences, the tempo of cultural evolution in man has accelerated exponentially. We must not forget, however, that language and especially writing are recent evolutionary creations in the long line of hominids. They are truly new, and exist only in man. Language and writing are products of biological evolution! Man's genetic endowment nearly always permits him to learn a spoken language, and in most cases to learn to read and write as well. Neither of these means of communication, however, appears spontaneously. Each infant must serve its apprenticeship. Conversely, one must note that no ape has learned an articulated and reflective language or, for that matter, writing and reading.

The interactions between biological and cultural evolution are double headed. The abilities man gains from cultural evolution feed back and modify his biological evolution and that of other organisms. The consequences of cultural evolution have modified the gene pools of the human species itself. Beyond his effects on his own species, man has exterminated many others. Still others he has modified by artificial selection and by inducing, or at least preserving, mutant forms. Man has brutally and dangerously modified environments, both his own and those of other species. Man's cultural evolution has had enormous repercussions for the biological evolution of all organisms; it seems unlikely that any species has escaped unscathed.

By virtue of his cultural evolution, man has become the first species capable of both his own destruction and that of large numbers of other species. Moreover, man is capable—at least in principle—of consciously guiding his own evolution. Even though a formal blueprint for human evolution has not yet been drafted, human beings can rapidly develop the means for directing both their own evolution and that of other organisms. Indeed, actions that stem from culture will influence man's biological nature whether we like it or not: whether or not man evolves is not man's choice; the choice is whether he will evolve under his own guidance or under conditions established by default.

With no reservation, we affirm that man is a product of the same evolutionary process that has created the diversity of all organisms. But, by successive steps occurring over hundreds of thousands of years, cultural evolution has taken hold so completely that *Homo sapiens* occupies, following the extermination of his predecessors, an extraordinarily predominant position among all organisms. Without recourse to any supernatural force and without drawing on any metaphysical interpretation, this unique evolutionary success can be explained through man's capacity to transmit all personal experiences by speech and writing from person to person, from population to population, and most importantly, from generation to generation. Man now accumulates knowledge incomparably more rapidly than does any other animal. This capacity, however, was acquired gradually, over an enormous period of time.

Language and writing have not existed since the beginnings of hominization; that must be insisted on once again. Although scientists are not absolutely sure about the origins of speech, it might be supposed that it initially consisted of only a few elements. Human languages have been enriched slowly, paralleling the development of a culture which demanded, in turn, the creation of an ever richer vocabulary. The development of language and of cultural knowledge has been interactive and interdependent: language and culture have coevolved. With respect to evolutionary change, the large differences existing between individuals of the same population today must be stressed. Robert (1953) has estimated that most French persons use fewer than 5,000 words. Certain rare individuals probably know more than 10,000 words, whereas a substantial fraction of the population does not use more than a few hundred. In part, such differences are determined by variation in the genetically determined capacity to learn. One might imagine that early in the history of human language only some individuals partook in its creation; they, in turn, could have been members of a single local population.

We do not know when man began creating an articulated language. Critchley (1960) thinks that in its creation, man had no need for an entirely new morphological apparatus. From an anatomical point of view, anthropoid apes should be able to speak. From anatomic, physiologic, psychologic, and cultural points of view, linguistic precursors must have existed in some animals even before man. Human language results from a coordination of factors that are found isolated among this or that lower animal. The transition from emitting animal sounds to speaking an articulated language was part of that evolutionary ensemble known as hominization. When instinctive and stereotyped responses were no longer adequate in human groups, more complex vocal reactions occurred. Reactions to personal contact plus differences between individuals were early keys to the socialization of man. The behavior of the individual members of a group had to be coordinated during the preparation for and carrying out of a hunt; in organizing the security of the group during the night; and for communicating knowledge to the young.

Critchley admits (as do others) that one cannot and may never be able to date the origin of language; nevertheless, one can speculate. Such speculation would take into consideration the volume and the form of the cranium; it would also recognize that the form of the jawbones determines the freedom of the tongue's movements. Murals created during the period of the Aurignacian, the Solutrean, and the Magdalenian cultures are scarcely conceivable without the beginnings of conceptual and symbolic thought; thus, language could have existed at that time. Reasoning in this manner, Critchley estimated that human language almost certainly existed twenty-five thousand years ago. On the basis of tools, arrowheads, and the utilization of fire by Neanderthal man, and even the flint tools of Pithecanthropus, Critchley suggests that language might even have arisen one hundred thousand years ago.

All animals that form societies—or even simple, nonstructured gregarious groups—communicate with one another; it is indispensable that they should. Such communication consists of signals, gestures, odors, songs, cries, and other attention getters. Mother quail can "tell" their little ones by a particular cry, "Go hide under cover," and by another sound, "Come back and warm yourselves under my feathers." Some animal signals persist in man. We use gestures, facial expressions, cries, and even roars with those with whom we would talk; these nonverbal signals constitute a repertoire that is used throughout the world, independent of the local oral language. Mother and child use such means of communication as long as the infant cannot understand the mother's spoken language. Communication of this sort is even possible between different species. A dog, a cat, or a performing tiger knows the meanings of certain gestures and sounds of man—and vice versa.

We agree with Simpson (1969), however, in refusing to use the term "language" for all these forms of communication. He does not, for example, speak of a language of the bees. Articulated language exists only in man; indeed, it is the most important diagnostic character permitting man to be distinguished from all other organisms. Other forms of communication do not permit discussion, abstraction, or symbolization. Furthermore, man's articulated language does not arise spontaneously; it must be learned by each individual. The capacity to learn one or more languages is genetically determined. Learning, consequently, requires a biological foundation. The capacity to learn even differs from one individual to another according to their genotypes. But that capacity exists only in man and, without doubt, in man only since a certain prehistoric epoch. Language appeared during hominization; its development represents the most important event in man's evolution, for it permitted his cultural evolution.

If chimpanzees, parrots, crows, myna birds, or even starlings are capable of pronouncing the words of human speech in different languages, and even with characteristic accents, that proves only that these animals are anatomically able to produce sounds resembling those uttered by man. None of these animals comprehends or uses a language for human-type communication. Even if these animals use words in the proper context, their "speech" does not constitute a reflective language. The phonetic apparatus exists in all the species mentioned, but they do not possess the necessary structure of the right cerebral hemisphere. The human brain permits all normal individuals of our species to learn at least one language. Spoken language is an evolutionary adaptation that is characteristic of man. To ask whether language represents the origin of cultural evolution or whether, to the contrary, cultural evolution created language is not a valid question: the two are parallel and independent phenomena that have coevolved through close interaction.

A few words may be said concerning writing and printing. These two means for preserving through time and for spreading through space the entire experience and thought of one individual have been and are extremely powerful accelerators of cultural evolution. Radio and television offer the hope of even more rapid progress in the cultural evolution of man but, at the same time, the appalling prospect of pernicious conditioning for large portions of the human population. The possibility of genocide or even of the destruction of our species is inherent in cultural evolution; cultural evolution provides the means for self-annihilation.

As a result of biological and cultural evolution, our species has attained the ability to destroy itself as well as many other species. To some extent, we are capable of controlling our own evolution. What use shall we make of that ability which of all living organisms we alone possess?

Some biologists think that evolution has ended, that we have reached the summit. That is a dangerous illusion. Technocrats, on the other hand, promise us ever more rapid progress leading to ever more marvelous tomorrows. That is an even more dangerous illusion.

Stopping evolution is impossible. All species have either become extinct or have evolved. Cultural evolution in no way guarantees an unlimited future for our species. Our knowledge of evolution says that the survival of a species depends on its ability to adapt to changes in both biotic and abiotic environments. For all other species, survival depends on the adaptation of gene pools through biological evolution. For man, biological evolution is inexorable. We are distinguished most from other organisms, however, by our cultural evolution, which has itself led us to the brink of disaster.

Our species contributes massively to the degradation of the world's biosphere. Even more dangerous, however, is the fact, unique among all organisms, that a tiny fraction of our species, a mere handful of individuals, dominates the others to an extraordinary degree. This power is sometimes ostentatious; more often and more dangerously; however, it is hidden. To be sure, hierarchies of dominance exist in other species. Cultural evolution, however, has given an incredibly large dimension to the position of dominance in the case of man. The power of dominant individuals and the means of their domination is such that the number of individuals they control greatly exceeds anything observed in other species. Needless to say, these dominant individuals do not always have intellectual and moral qualities commensurate with their positions.

Natural selection among other animals has favored the development of behaviors that impede abuse by dominant individuals. Dogs, wolves, foxes, other mammals, and birds furnish examples of behaviors which without doubt are genetically determined and which permit the establishment of hierarchal social arrangements by mutual agreement.

When a wolf accepts an inferior position in a pack's social order, he presents his neck to his antagonist, thus avoiding a fight. The dominant individual does not attack another who signals his submission. To invite a fight, a dog places his muzzle under that of another; otherwise, he lies on his side to show an absence of aggressive tendencies. Conversely, the individual wolf or dog that has decided to battle for a dominant position scarcely ever attacks without warning. He clearly manifests his intention by postures, facial expressions, and significant positions of his ears, tail, and mane.

Among men, the creation of enormously destructive arms and the possession of dreadful power by those in dominant positions have outstripped bio-logically based behaviors which would avoid abuse. The danger to all persons arising from this imbalance is a consequence of cultural evolution. Progress in cultural evolution itself is needed to eliminate this grave danger. No more hope is to be found in biological evolution—for example, by natural selection favoring altruistic genotypes. Natural selection cannot act with sufficient speed to save our species.

Vestiges of a regulated cultural evolution exist. One might cite the rules of chivalry of the Middle Ages. The fundamental English law of Habeas Corpus of 1679 citing the Magna Charta of 1215, then the Declaration of Indpendence of the United States in 1776, the Declaration of the Rights of Man and the Citizen in 1789, the action of Jean Henri Dunant, founder of the Red Cross, and the 1948 International Declaration of the Rights of Man are important milestones marking progress in the legal basis of informed consent—progress made necessary by the frightful possibilities of destruction that cultural evolution has given man.

That cultural evolution has endowed man with science, medicine, technology, means of transport, printing, literature, art, music, and an extraordinary enrichment of inner and outer life is not in question. For these gifts we are prepared to use the term "progressive evolution," although all value judgments should be used with caution in the domain of evolution.

But cultural evolution has also engendered immense danger; for example, despite the efforts of many the menace of a nuclear war still persists. The installation of numerous nuclear power stations has, and will have in the future, unfortunate consequences for man and other species. Pollution of the air, water, and land will create intolerable conditions for both man and other species if Draconian measures are not soon taken. Even now, most food and drink is purposely contaminated by soluble coloring agents, sweeteners, stabilizers, and even hormones. Many of these substances have proven to be teratogenic, ulcerative, carcinogenic, or mutagenic. The rising number of persons on earth, also a consequence of cultural evolution, aggravates all other problems.

An additional consequence of recent cultural evolution, and one boding ill for man's future, is standardization—the homogenization and impoverishment of imagination that is imposed by the press, radio, and television. The situation will worsen when TV programs are relayed by communications satellites to the entire world. The consequence will be a stultifying loss of imaginative intellectual forces. We will witness an abasement of man's culture because standardization never seeks a high level—on the contrary, it always falls to the lowest common denominator.

Intellectual pollution is as dangerous as pollution caused by radioactive and toxic substances. Cultural evolution must be rich, heterogeneous, multiform, and open to all new ideas and new forms of expression. Fortunately for the future culture of humanity, there do exist those who march to a different drummer; they include an increasing fraction of the young who oppose intellectual and other forms of pollution. Not surprisingly, they rebel occasionally against technology and science. They represent the best hope for the rectification of today's unhappy situation. They will learn, we hope, to distinguish between good and bad technology. The human species must become more humane; that change can, itself, be regarded as an evolutionary one.

Summary

Upon leaving the prehominids, biological evolution created among hominids the necessary conditions for a new form of evolution, cultural evolution. Man has now arrived at a stage which gives him extraordinary powers over his own evolution. Neither stagnation nor regression is possible. The evolution of man may be arrested by collective suicide: suddenly by an atomic war or slowly by pollution, anoxia, and poisoning. These possibilities are not fictional; on the contrary, they are based on cold logic.

Other organisms undergo biological evolution according to laws and mechanisms which they do not understand; man alone understands evolution better and better—both his own and that of other species. Not only is he able to act and to intervene in his evolution, but *he does so*, massively, for better or for worse—whether or not he wishes to do so. For this reason, research on evolution, and on the evolution of man in particular, is of utmost importance.

The cultural evolution of man must now create a truly humanistic ethic. The survival of the species will be possible only if we dispose of ethics and altruisms that are limited to family, tribe, or those who hold the same religion or philosophy.

The menace of nuclear war, atmospheric pollution, radioactive pollution, and the pollution of the oceans concerns all peoples. The danger that our species will disappear can be banished only by constant effort; humankind's existence requires a worldly, humanistic ethic.

The situation of our species is such that scientists and philosophers cannot afford the luxury of holding purely academic discussions on the question of evolution. It is no longer a question of knowing who is right: the creationists or the evolutionists, the Lamarckians or the Darwinians, the vitalists or the materialists. The questions "Where have we come from?" and "Where are we?" are passé. Today, the question inscribed by Gauguin on one of his most beautiful paintings of Tahiti is the most important of all: "Where are we going?"

Thus, the controversy over the origin of man is not the academic matter some biologists and philosophers believe it to be. On the contrary, it is a matter involving all humanity. Consequently, the evolutionary biologist has acquired a great responsibility for his fellow man.

For the creationist, the species has been created once, for all time. For him, it is enough for mankind to remain on (or return to) the right track. Brotherhood has been bestowed on mankind from the beginning; it is enough to win back once more those who have forgotten their religion under evil influences. Whatever may happen, God wills it.

For the vitalist and all those who postulate unknown, and even unknowable, evolutionary forces, we can only passively evolve because the mechanisms of evolution are beyond our comprehension.

For the evolutionist who thinks that we already know a certain number of evolutionary factors or mechanisms, the responsibility for mankind is large. Because man can guide his own evolution to some extent, the evolutionist has the duty to foresee and recommend possible future paths. We are no longer animals who submit passively to the laws of nature. We have foresight and can act accordingly. Certain sorcerer's apprentices know this and do act with limited foresight, but for their own profit rather than the collective good.

The thorough study of the biological and cultural evolution of man and a knowledge of past evolutionary tendencies of hominization may help us find paths leading to the future *humanization* of our species.

CHAPTER 6

The Biological Bases and Evolutionary Functions of Esthetic Sensations

Having discussed man's cultural evolution in chapter 5, we shall now turn specifically to the evolution of art, a subject that is both complex and controversial. Can one really treat art as an aspect of evolution? We believe so. The question, we must emphasize, is not the *history* of art; in the following pages, we shall search for an adaptive role for art (or esthetics generally) in human evolution.

To a greater extent than usual, various definitions of "art" and "esthetics" must be examined; otherwise, our discussion will be garbled beyond comprehension. The *Encyclopedia* or the *Descriptive Dictionary of Sciences, Arts, and Crafts* of Diderot and D'Alembert (1751) adopted a broad definition: "One generally gives the name 'science' or 'art' or 'discipline' to the hub or point of juncture at which are joined past observations, in order to form a system. . . ." They present some examples: Grammar, mechanical arts, liberal arts, philosophy, the art of glassmaking, and angelic art. Painting, sculpture, and similar items, however, do not come under the name of art. The *Encyclopedia* says, "Man is only the minister or the interpreter of nature: he neither understands nor does he do other than what he knows of either by experiment or by reflection of things surrounding him."

Under "artisan" we read, "One says of a good shoe-maker that he is a good artisan; and of a clever watch-maker, that he is a great artist."

According to the *Encyclopedia*, the arts are capacities of man (which distinguish him from animals) for imparting through the aid of instruments and rules "certain definite forms on a base given by nature; and that base is either material, spiritual, some function of the soul, or some production of nature." The *Encyclopedia* distinguishes between the sciences, which invent, and the arts, which apply and produce: "Thus, one says of a chemist who is able to execute skillfully the procedures which others have invented, that he is a good *artist.*"

In the same *Encyclopedia*, the painter is repeatedly defined as an "artist who is able to represent all sorts of objects with the aid of paints and brush." For this chapter on the evolution of art, the following sentence from the *Encyclopedia* has special interest: "Experience amply proves that all men are not born with the same talent for rendering paintings."

The *Encyclopedia* affirms "that objects placed before the eyes of man seem to invite imitation. One should place painting among things purely agreeable since that art is unrelated to those which one accurately calls the necessities of life. [But] through all time those who have governed peoples have always made use of paintings and statues in order to better instill in their subjects sentiments in either religion or politics which they wish them to have."

The *Larousse* of 1866 derived "art" from the Sanskrit *kri*, or *kar*, which signifies "to make" and which has given the Latin *cre-are*. The first meaning is thus: manual skill industry. The *Larousse* recalls that this meaning persists in *artist* and *artisan*.

For Bossuet, art is the embellishment of nature. For Vanvenargues, natural endowments are worth more than those of art. For Grimm, art is to nature as a beautiful statue is to a beautiful person. And for Balzac, the greatest efforts of art are always a forgery of nature. According to Bacon, art is man added to nature. La Bruyère affirms that art sometimes impairs nature in attempting to perfect it. And Victor Hugo says that "art is a creation peculiar to man. Art is to man as nature is to God."

Under this general view, however, certain writers see art forms among animals. La Fontaine says of beavers that they build their houses and cross ponds over bridges which they themslves have built with considerable skill. In Fontenelle one reads: "Most animal species such as bees, spiders, and beavers have a particular art, unique, and which not one among them has first invented; man has an infinity of different arts, with which no person is born, and whose glory belongs to individual persons." We might recall here that Lamarck made a sharp distinction between the reality of individual organisms (which are products of nature) and the categories of taxonomic classification (which are part of art).

Usually, one contrasts art with nature. One calls *artificial* that which is produced by art, as opposed to *natural*. Today, the term "art" means esthetic productions, notably paintings. Taine asked himself about its goals: "The conclusion still seems to be that eyes have remained fixed on nature in order to imitate it as closely as possible, and that art consists entirely of an exact and complete imitation." However, Musset complained that "art is constantly below nature, especially when it strives to embellish it." Balzac responds, "The goal of art is not to copy nature, but to express it." Cocteau says, even more simply, "Art exists at the minute when the artist deviates from nature."

Here is the concept of art proposed in 1818 by Victor Cousin: "Religion is needed for religion, ethics for ethics, and art for art." From that definition arises the idea that art should be useless. Th. Gautier said fervently, "Nothing that is beautiful is indispensable to life . . . there is nothing truly beautiful which has a purpose; everything useful is ugly." Uselessness, however, is not sufficient to guarantee beauty. Stendhal thinks that "all that which is extremely beautiful, in nature as in the arts, is a reminder of what we love. . . ."

No objective criterion concerning the esthetic quality of works of art yet exists. One loves or one does not love a work of art, according to personal tastes. Montaigne has presented this point of view: "It is probable that we scarcely know what is beauty in nature and in general, since both man and beauty come in such diverse forms." Similarly, Pascal said, "Fashion and the country determine what is called beauty."

This rapid survey of some opinions on art demonstrates the absence of any consensus. In his discourse on esthetics delivered at the Second International Congress of Esthetics, Sciences, and Art in 1937, Paul Valéry (1968) introduced himself as an amateur, very much embarrassed for appearing before the most eminent representatives of esthetics. Being "a man sufficiently foreign to that science," however, he went directly to the heart of the matter:

"I tell you from the outset that the very word *esthetics* has always truly amazed me. . . . It causes me to waver between a strangely seductive idea of a science of beauty which, for one thing, would cause us to identify with certainty between that which we love, that which should be hated, that which should be acclaimed, and that which should be destroyed; and which, on the other hand, would teach us to produce with certainty those works of art of undoubted value; and, in reference to this first idea, another—a science of sensations—no less seductive. . . ."

In effect, a science of beauty would assure us concerning value judgments, while the science of sensations would teach us their physiological bases. Valéry asked himself about the biological bases of beauty. Is that what causes pleasure? If so, it would be "a purely sensory act [which] has easily enough accepted an honorable and limited functional role: one assigns to it a generally useful function under the mechanism of preserving the individual, and in complete confidence through that to the propagation of the race—"

"But," he continued, "there are pleasures and pleasures . . . there are those with no purpose in the economy of life. . . . Pleasure, finally, exists only momentarily, and nothing is more personal, more uncertain, and more incommunicable. . . . The only purpose of a science of beauty seems to be demolished fatally by the diversity of beautiful products that are acknowledged in the world and in time. . . . Individuals amuse themselves as best they can, and the malice of sensibility is infinite."

Valéry denies a dogmatic and infallible esthetics capable of decreeing and judging logically what is or is not beautiful. He does not ignore the importance of a rational esthetics. He does not believe, however, that it should dominate taste, impose itself on artists and public alike, or force people to love that which they do not love and abhor that which they love.

Among artists, art critics, theoreticians of esthetics, and even amateur consumers of art, one finds contradictory personalities. Some are endowed with a fixed mentality that bolsters them while they adhere to an academically established code of beauty. For many, the highest ideal is the art of the Hellenistic Greeks. At the other extreme are those who reject all criteria and all value judgments. For them, each work has its intrinsic value; everything is incidental to the esthetic sensations each work of art evokes from the observer.

The idea of an evolution of art and of esthetic sensations is excluded by both of these extreme positions. We shall propose in opposition to these extremes an evolutionary conception of art. Our argument could be applied to all esthetic expressions, but we shall consider only pictorial art.

Our goal, as we said earlier, is not to present a history of art, but to inquire into the biological bases of esthetic expressions and perceptions. Are they gratuitous and useless? Or do they have biological functions in man and other organisms? Are there evolutionary tendencies in esthetic productions and perceptions that are based in biology? If so, is there an evolutionary progress in the arts? Perhaps one can establish objective criteria by starting from an evolutionary concept of art.

What we propose to undertake is complex and controversial; an evolutionary basis for esthetics is an idea practically unexplored. Nevertheless, certain elements of human esthetics have vestiges that are found among other organisms. Consequently, it is reasonable to discuss art under the aspect of biological evolution.

In the usual sense of the word, art, or the arts, is a human phenomenon. Animals do not produce works of art. The capacity to communicate with other individuals by an artistic expression composed, structured, freely chosen, and reflective is one of the characteristics which distinguishes man from other organisms. That need not signify, however, that any factor, any element, any capacity, whether sensory, intellectual, or psychic, or any motivation and stimulus that intervenes in the esthetic expressions of man does not exist in some form and to some degree in other organisms.

The capacity of man to create works of art has not appeared suddenly, full grown, from nothing. That capacity was not given to man directly by a Supreme Being who had anticipated endowing man with this special means of communication.

Artistic expressions have not always existed in man. They were formed by degrees in the course of humanization. The arts are a product of man's cultural evolution, which is itself a consequence of biological evolution. Consequently, it seems reasonable to search for the biological bases of art. It will not be easy. To place in evidence the whole evolutionary process in precise detail will not be possible. Of that, there is no doubt! But is it possible to do so in other evolutionary matters? We know only the smallest bit of the entire history of any species, of organisms or their anatomy and physiology, and of their behavior. Nevertheless, few would totally deny an interest in research on biological evolution or despair of gaining thereby at least some understanding of its workings.

Not only shall we try to understand the evolution of esthetics, but we shall also try to profit from our efforts in order to better understand the place and the function of art in human populations. We may even arrive at certain criteria for the appreciation, or at least the understanding, of art, knowing full well that entertaining such a hope may startle most persons.

Biological Functions of Forms, Patterns, and Colors in Plants and Animals

Most persons derive pleasure from contemplating a flower, a butterfly, a bird, or a gazelle. Are these forms, patterns, and colors that we find so beautiful simply tricks of nature, without functional value? Or do they have a biological role that could have been established by evolution? Darwin (1871) claims the latter in his theory of sexual selection—at least for certain decorative characters which one finds among animals. He thinks that, even more than weapons that males use in obtaining mates, all sorts of ornaments and sound-producing organs have evolved under the pressure of sexual selection.

Darwin argues that secondary sexual characteristics are variable. This variation permits animal breeders to modify these characters according to their tastes; in accord with prevailing standards of beauty, Selbright Bantam cocks, for example, have come to possess new and elegant plumage and a characteristically upright carriage. Under natural conditions, females similarly augment the beauty of males by a selective preference of the most attractive ones. Darwin added that this implies on the part of females a capacity of discrimination and the existence of a certain taste—characteristics that seem extremely improbable at first sight.

Among the European sandpipers (*Philomachus pugnax*), Selous (1929) observed the nuptial dances which males perform en masse before the females on assembly grounds. He claims that females mate preferentially with males who are more vividly colored; other, drabber males do not participate in the perpetuation of the species.

Among the heath cocks (*Tetrao urogallus* and *Lyrurus tetrix*), the males assemble during February and March in arenas (which are reoccupied annually) in order to perform their nuptial parades. The males use this opportunity to present their skill and the colors of their plumage. Each male possesses a field of tournament, his private territory which he defends against intrusion. During the nuptial parade the little heath cock raises his tail and displays it in the shape of a lyre. He shows his bluish-black feathers and exposes his white underfeathers. His red wattles and comb are visibly swollen. From time to time the male leaps upward, agitates his wings, and exhibits the transverse white band as well as the bright colors of his underwing. During egg laying the females take an interest in the tournament of males and promenade on the periphery of the arena. When a female approaches a parading male, he either crouches on the ground before her or begins to run a few steps about her while extending his head horizontally. When the female reacts by crouching down and effecting sinuous movements with her neck, the male stops his running and covers the female (Verheyen 1950).

Even among certain species of *Drosophila* endemic to the Hawaiian Islands, males occupy and defend territories on ferns or the trunks of trees. There they perform their nuptial acts. Females join males within their territories in order to be fertilized (Spieth 1968). In many Hawaiian species, such as *D. grimshawi* and *D. comatifermora*, one sees a pronounced sexual dimorphism, whereas such dimorphism is nearly lacking in other *Drosophila* species. These observations constitute a strong argument for the Darwinian theory of the evolution of secondary sexual characters by sexual selection. For many endemic Hawaiian species of *Drosophila*, even qualified taxonomists are not able to distinguish the females of different species, whereas the males of these same species are notably different because of their dimorphism and their sexual behavior.

For most species of birds, the males' nuptial parade offers an opportunity to expose the particularities of their plumage. We might recall the spectacular parade of the peacock, and the marvelous plumages of the bird of paradise and the hummingbird.

In other classes of animals the nuptial display of the males also exposes patterns or vivid colors for the stimulation of the females. Often during the period of reproduction the coats or the plumages of males exhibit colors more vivid than are characteristic during the rest of the year. The alpine triton (*Triturus alpestris*) wears an elegant nuptial habit: a nearly black body with lilac reflections and an ornamented, scalloped dorsal crest of a vivid yellow. The flanks and the sides of the head exhibit a silver or golden band and another that is sky blue; above these bands are contrasting large black spots. The belly and throat are of a vivid red-orange. After the nuptial and aquatic stage, this same triton has a dull blackish skin that is yellow on the underside.

All colors and designs which we regard as elegant and pleasant need not

have a role in sexual selection. Darwin himself mentioned marine ane-
mones, some medusas, siphonophores, ascidians, and other colorful animals among whom sexual selection is not at all evident. Certain colors offer
mimetic protection, others are simply the result of the chemical nature of a
tissue. Darwin (1871) gives this example: "There is scarcely a color more
agreeable than that of arterial blood; but there is no reason to think that the
color of blood has of itself any advantage; and although it augments the
beauty of the cheeks of young girls, no one pretends that the color of blood
has been acquired for a purpose."

Colors of the autumn leaves in American forests are notoriously splendid;
nevertheless, no one attributes to them a selective advantage. Nor to the
marvelous colors of hermaphroditic nudibranchs whose biliary vesicles can
be seen through their translucent skin. We shall discuss these aspects of the
beauty of nature somewhat later. Before that, however, we shall mention
briefly three situations where colors and patterns have indisputable biological roles.

Among plants, besides asexual modes of reproduction, different forms of
sexual reproduction exist. When fertilization is accomplished by the aid of
water, as in the bryophytes and the pteridophytes, or of wind, as among
conifers, flowers are normally subdued. Among the majority of phanerogam
species, pollination is ensured by animals, mainly insects. Among these plants
one finds an extraordinary variety of forms, colors, patterns, and odors of
flowers. The role of insects as pollinators was described in 1793 by Sprengel.
More recently the work of von Frisch (1914) and others has shown that bees
and other insects (for example, the dipteran *Bombylius fuliginosus*) distinguish colors very nearly as we do, except that the spectrum of light visible
to them is lowered toward the ultraviolet region. For certain plants, birds
(in America, the hummingbirds; in southern Europe, the Meliphagidae and
the Nectarinidae) ensure pollination. A particularly interesting situation is
that of the complex of *Ophrys scolopax* and some other orchids. Each plant
produces a pheromone that specifically attracts males of certain species of
Hymenoptera. Because the labellum of the flower imitates the form of a
female hymenopteran, the male attempts to copulate; in doing so, he picks
up pollen which he transports to the next flower he visits.

Homochromy and protective mimicry offer still other examples of the
biological significance of forms, colors, and patterns. The industrial melanism of *Biston betularia* (and of certain other species of Lepidoptera) protects

the dark forms against predatory birds in the industrial regions of Britain and continental Europe.

Certain fish modify their color and design by means of chromatophores in order to match the background on which they rest. The chromatic camouflage of the chameleon has become legendary. The mimetic coloration of cuckoo eggs matches that of the eggs of the host species—a species chosen by the female cuckoo so that she can lay an egg in the host's nest. There is often, if not always, a marked similarity between the eggs of the host and that of the cuckoo. Imprinting while being raised by adoptive parents conditions the young female cuckoo to the appearance of her adoptive parents; she later seeks out such nesting birds when she searches for a nest in which to lay one of her own eggs.

Here we can recall mimicry, which is frequent among insects, and which offers protection through the imitation of inedible species—species that are toxic or possess stings. The butterfly *Papilio dardanus* mimics in different parts of Africa various species of inedible butterflies; such mimicry protects *Papilio* from predators (Ford 1971). The mimetic resemblance is in many cases extremely precise. *Papilio dardanus* copies the colors and color patterns of wings, as well as the overall behavior of its models. Beetles mimic ants; lepidopterans mimic bees. In every known case, shapes, patterns, and colors are important in providing protection against predators, especially birds.

In the present context, the moth *Zygaena ephialtes* is particularly interesting. For predacious birds this insect has a repugnant taste. It possesses diverse forms and colors which comprise a Müllerian system of mimicry (Bullini et al. 1969): yellow forms of *Z. ephialtes* belong to a mimetic chain which includes the lepidopteran species *Amata phegea*, *A. marjana*, and *A. ragazzi*. Where the species of the genus *Amata* are rare or absent, one finds red forms of *Z. ephialtes* which are part of a chain of mimetics that include phenotypes resembling other species of the genus *Zygaena*, but also of other insects—for example, the beetles Cerambycidae, Meloidae, Cleridae; the lepidopterans Arctidae; and the homopterans Cercopidae.

The protection that toxic and repulsive substances provide the *Zygaenae* against predators is reinforced by the mimetic coloration which causes them to resemble still other toxic insects. Predatory birds distinguish between butterflies according to their form and color. They quickly and efficiently learn to shun repugnant butterflies when they find themselves confronting similar forms.

Bullini et al. (1969) state that the capacity for distinguishing different insects varies among different species of birds. Starlings and blackbirds are better entomologists than are nightingales. For our argument, it is important to note these differences in ability with respect to visual distinctions.

We have concentrated here on the biological functions of shapes, patterns, and colors. A similar discussion could have been made for sounds, or even for bodily expressions, gestures, and those postures which express themselves in man by song, music, sculptures, and dance.

We do not pretend that an esthetic sense comparable to man's permits birds, butterflies, and many other organisms to develop those phenotypes which, in fact, do bring us an esthetic pleasure. Later, we shall see that certain animals do possess a small capacity for esthetic appreciation. Such a capacity is excluded, of course, for plants, because they have no organs for the perception of colors. Butterflies may or may not have this capacity; it is reasonable to believe, however, that birds and mammals actually do have it. If we argue that higher vertebrates exhibit elements of perception and that they appreciate esthetic harmonies, we do not necessarily claim that these animals also create works of art. Among those persons who have never created a work of art, the capacity to appreciate art extends from an almost total absence of esthetic emotions to extremely intense ones. Furthermore, an individual who is extremely sensitive to figurative art may be totally insensitive to artistic musical expressions.

In brief, granted that man alone produces works of art in the usual sense of the word, it is nevertheless unrealistic to claim that an absolute difference separates man and other organisms when art is a question of perceptions or even of esthetic sensations. Forms, colors, patterns, and bodily expressions play an important biological role in the great majority of organisms. From the absence of the perception of forms and colors among plants and among primitive animals, one passes to more and more advanced means of perception. This is related to sensory organs. Man does not possess the most advanced sensory organs of any sort. But "progressive" evolution has led to a central nervous system which alone is capable of integrating sensorial perceptions, thereby producing an esthetic emotion. From this point of view, man must represent the highest stage of evolution. Human art, however, did not appear suddenly, full blown. There are forms, colors, and patterns in the whole world, not just among living beings. They play a role among organisms which do not themselves perceive them. Biological evolution has led to capacities of perception because they offer selective advantages. And it is on the large biological base ever present in nature that the development of the human brain has permitted (and still permits) a tiny minority of human beings (representing, without doubt, particular genotypes) to create works of art.

Is There an Esthetics Among Animals?

One can distinguish, on the one hand, between works of art and useful objects manufactured (in the old and pleasant sense of that word) by artisans and, on the other hand, the attitude of individuals who contemplate works of art or useful products. There is, however, no clear-cut difference between a useful or even necessary object and a work of art. In going through museums, antique shops, works of art, and catalogues, it may perhaps be possible to classify certain items this way; others, though, cannot be classified.

While writing these lines, I [E. B.] have before me an eighteenth-century ivory which represents a young woman covered only by a ribbon. It is a very beautiful sculpture: surely, a work of art. My little statue was manufactured, however, by an artisan for a Chinese doctor, who asked his female patients to point out on the statue where they suffered. A bit farther away, on a cabinet, is a square Japanese flask which once contained sake; on its four faces are cranes in a tremendously elegant flight, made even more beautiful by the subtlety of almost imperceptible distant clouds, and by the cracks in the enamel that were made purposely by a fine craftsman. It is evidently an artisan's production and a useful one. Further, on the wall is a gouache by Hans Erni that represents a couple. It is not a useful object, at least at first sight. It is undoubtedly a work of art, created by an artist. But the couple whom I contemplate from time to time are not useless. Despite all the psychic aggressions of the real world, they help me recreate an equilibrium that is indispensable for my well-being and for my work as a scientist. Furthermore, Hans has often told me that what drives him and enables him to create works of art is the need and the satisfaction of working with tools and materials; his drive is the creative art of an artisan.

Together with Professor Dobzhansky, I have seen numerous works that are classified without hesitation as great art: the paintings of Giotto at Padua and Pompasa and of El Greco at Toledo, the marvelous Byzantine mosaics

at Ferrara, the stalwart capitals of the cloisters of Elne de St. Martin de
Cuxa, and much more. These are not works of art for art's sake; they are
not "useful works" in the normal sense, but neither are they useless things.
These admirable works, in our opinion some of the leading works of art,
have had (and still have today) a religious function—consequently, a function that is human and social.

To distinguish between a work of art and the product of an artisan who
has an esthetic capability is impossible. All objects that are produced with
an esthetic sense have an artistic value. But what is the criterion of an
esthetic taste? We cannot provide a definite answer to that question; to do
so is impossible. Human beings, as biological organisms, are variable just
like all other living creatures. This is true with respect to both sensory perceptions and the cerebral integration which results in an esthetic pleasure.
Intervening factors are so numerous, and the interactions between them are
so complex, that each individual, in effect, has his own esthetics.

For that reason, we think that one can also speak of an esthetic in certain
higher animals, and of an evolution of esthetics. We must carefully avoid
being anthropocentric. The question is not one of projecting our own sense
of esthetics onto animals. On the other hand, the experimental analysis of
esthetic sensations by objective methods is extremely difficult because the
problem is so complex.

We might recall, however, that sexual selection exists in many animals;
matings scarcely ever occur by chance. Furthermore, among birds and
mammals, colors and patterns have a part in the choice of partners. It is not
anthropocentric, then, to attribute to higher vertebrates a degree of esthetic
sense. The perception of forms, colors, and designs is among most higher
animals as fine as our own. Most likely, however, only higher vertebrates
are gifted with a capacity for cerebral integration that permits them to sense
an esthetic harmony and to make a more or less conscious choice in the
case of sexual selection. Although we cannot know how a bird or a mammal
"feels" beauty, the anatomy of the brain, and above all of the telencephelon
and its cortex, allows one to believe that the degree of intregration of the
perception, not only of the elements (colors and others), but of the totality
which makes an esthetic harmony should be much less developed in birds
and lower mammals than it is in man. In thinking of evolution, however,
one should not forget that in our own species the range of capacities and of
intensities of the esthetic sense is largely exposed: it is lacking completely in
some individuals but is extremely refined in other (rare) persons.

A possible method for detecting discrimination of an esthetic sort has been used by Rensch (1965). Tests made on persons showed that they preferred rhythmic and symmetrical designs over similar figures with irregularities, as long as the choice was not determined by cultural influences. Without doubt, said Rensch, it is easier to apprehend repetitions of rhythmic and symmetrical figures than a design with irregular lines. The facilitation of perception by regularity is felt positively; therefore, it produces an esthetic pleasure.

Apes and crows love to play with small objects—for example, pieces of cardboard that exhibit rhythmic designs. Pablo, a capuchin monkey, when offered a variety of pieces, examined all of them before choosing one, two, or three for playing. Numerous tests have proven Pablo's preference for regular designs. Analogous experiments with a rook and with some jackdaws have also shown their preference for regular designs.

A collaborator of Rensch, Tigges, noted that jackdaws prefer pure colors to blended ones. We have made a similar observation with a crested tit that tore up little bits of paper placed at the bottom of her cage, which she then held in her beak. Upon presenting her with four kinds of paper at a time, we noted that she clearly preferred pure colors to blended ones. Between a sharp red and a dull red, she most often chose the first. She also clearly preferred red and green and ignored yellow. Moreover, we noted that this tit temporarily preferred a novel color which she had not previously encountered.

Among most individuals of three species of fish, Rensch (1965) also noted a choice between regular and irregular forms, with a clear preference for the latter. That all tested animals systematically chose between different designs shows an elementary esthetic sense.

Among the bowerbirds of Australia and New Guinea, the Ptilonorhynchidae, one finds not only a choice of objects of different colors but also an active expression of esthetic sentiments. Under the influence of sexual hormones—that is, before and during the mating period—the males perform displays, each on his own territory. These males build (depending on the species) different types of hedgerows, bowers, or huts (Marshall 1954). Certain species place colored objects before or within the huts—for example, shells, berries, flowers, fruits, and feathers. They pluck fresh items to replace wilted flowers, rotted fruits, and discolored or broken feathers. Normally, members of one species always choose the same color(s).

Our friend Ford has observed some bowerbirds in their natural habitat.

He placed in front of a bower some pieces of cardboard of a color very different from those found among the decorations arranged by the male. When the bird returned, he gave some strident cries and angrily tossed out Ford's cardboard pieces.

The decorations of the bowerbirds' huts have been attributed alternatively to an esthetic motivation or to a purely playful activity. Gould (1865) studied the behavior of these birds. He describes the richness of the decoration and notes that in certain cases some shells were sought by birds over distances of several kilometers. He supposes that these bowers are places for nuptial displays by males where they encounter females. Brehm (1866) gives pictures of the huts of two species, of which those of *Chlamydera maculata* were also reproduced by Darwin (1871), and mentions the opinion of Gould concerning the function of the constructions in the reproductive cycle. Darwin (1871) claims that the bowers (which are built on the ground) are constructed for the sole purpose of the nuptial display, because nests are built in trees.

Marshall (1954), who has made a profound study of the behavior of bowerbirds under both natural and experimental conditions, concludes that the construction of their decorated bowers is connected to sexual activities: "These complex and remarkable phenomena are probably the expressions of innate behavior patterns." We agree with Marshall when he refuses to consider these extraordinary constructions as having been decorated exclusively for pleasure and as an expression of an artistic and esthetic sense on the part of bowerbirds. He opposes Romanes (1892), who postulated that Darwin's theory was not able to explain esthetic phenomena such as those of bowerbirds, because they have no survival value. For Romanes, bowerbirds are accomplished artists. Marshall rightly insists on the utility of this behavior. Indeed, sexual selection provides an understanding for its evolution. In saying that, Marshall adds that given the present state of our knowledge we can neither deny that bowerbirds have an esthetic sense nor prove that they have. Some species of bowerbirds choose objects which to us appear beautiful. That does not prove, however, that these objects have a similar esthetic effect on the bird. The birds choose as a result of their genotype and physiology; the choice may well be mechanical.

If all bowerbirds made collections of only whitened bones and empty snail shells, as does *Chlamydera nuchalis*, one would insist less on their esthetic sense. But these piles of bones and shells are not less beautiful for that species than are the magnificent displays of blue and red blossoms of *Chlamydera lauterbachi* for it. Marshall states, "It would evidently be inconceivable to suggest that bowerbirds—or all other similar birds—do not sense pleasure in performing their vocal and architectural activities. But whether that pleasure has much in common with that which is felt by man in comparable activities remains to be proven."

We also agree with Marshall when he says, "We are probably able to understand something concerning the origin of certain behaviors of man by the study of innate reactions (not conditioned ones) of bowerbirds, but it is dangerous to project in an anthropomorphic manner certain qualities of mankind onto bowerbirds, qualities that require a strongly developed cerebral cortex. . . . Whereas certain authors have tried through a lack of critical sense to invest bowerbirds with human qualities, it is also adequate to speak generally of a certain bowerbird-ness in mankind."

Even without having performed the experimental studies demanded by Marshall, it is obvious that under no circumstances can birds have esthetic sensations comparable to those of man. Important differences in the structure of the telencephalon and, further, the essential fact of man's cultural evolution permit us to make this claim without hesitation. The esthetic sense of modern man, incomparably more evolved than that of other animals, has emerged evolutionarily from that of the primitive beings who were our ancestors. In passing, it should be noted that the case of an "esthetic production" such as that of bowerbirds is exceptional among animals.

A quite different situation is that of the higher apes, who, under experimental conditions, produce drawings and paintings. This cannot reflect a fixed or hereditary behavior, because apes do not paint in the wild. Even under experimental conditions, apes do not draw spontaneously. They must be stimulated by at least one demonstration.

For more than twenty years, many experimenters such as Morris (1962), Rensch (1965), and others have induced higher apes to create drawings and paintings. Although they do not reflect an evolutionary transition from animal to human art, such experiments are useful for comprehending the evolutionary progress toward human art. Rensch (1965) has clearly expressed this point of view in his article on the esthetic factors in higher animals.

Some chimpanzees, some gorillas, some orangutans, and some capuchins—when furnished with materials following a demonstration—produce (without an extended apprenticeship) designs and nonfigurative paintings. Is it only play, a pleasant activity without significance? Or, to the contrary, must one admit that these apes do not paint at random, that they ponder and then voluntarily produce certain patterns and designs?

These questions were discussed animatedly on the occasion of an exposition organized in London in 1957 by the Institute of Modern Art in which the works of two chimpanzees were shown. Morris, who had obtained these paintings by chimpanzees from the Zoological Garden of London, thought that one could detect esthetic factors—albeit limited ones—among higher apes. In his inaugural speech for the exhibition, Sir Julian Huxley claimed that the exhibited works were not simple products of play—that on the contrary, they raise fundamental questions concerning the development of human art.

Rensch has experimented since 1954 with a capuchin monkey (*Cebus apella*) named Pablo. He gave him chalk and showed him how to mark the wall of his cage by drawing the chalk over the black surface. Without hesitation, Pablo drew some lines and scribbles. Later, spontaneously and without help from the experimenters, Pablo produced clusters of lines and strokes with crayons and brushes. Frequently he creates fan-shaped designs.

Rensch (1957) has also experimented with three chimpanzees. Like Morris and Huxley, he has the impression that drawings made by apes are not chance products but that a trace of esthetic pleasure is involved. If painting for an ape is play, it is play determined by the joy of producing visible traces on a wall.

Rensch notes that Pablo occasionally produced some disorganized scribbles compared to those of an infant aged eighteen months. Nevertheless, his paintings almost always show a certain rhythm and a more or less harmonious composition. Among capuchin monkeys and chimpanzees, Rensch noted a detectable progress during the course of some experiments. At the start, they produced some crooked lines, apparently by chance, or again, they covered a large part of the paper uniformly with paint. Becoming more daring, the animals arrived at the creation of rhythmic designs—notably fan-shaped ones.

If Rensch gave the apes different colored chalks in sequence, they often produced nonfigurative paintings exhibiting a certain esthetic attraction. To be sure, they had not chosen the paper and the colors, but it must be admitted, says Rensch (1957), that his capuchin monkey and the chimpanzees are able to fill the paper harmoniously with lines and designs.

Obviously, to know if apes are motivated in their artistic activity by an esthetic pleasure is not easy. But the principle of centralization, noted frequently by Rensch, is an indication of the satisfaction experienced by an ape as the result of a certain harmony of orderly strokes. One often sees a certain equilibrium between the left and right sides of the sheet.

When Rensch gave his apes triangles, squares, and spirals of white paper glued on gray cardboard, they often followed, more or less, the white traces with their paint brushes. Huxley and Ford have observed a gorilla at the London Zoological Garden who several times carefully followed the contours of his shadow on the wall of the cage with his index finger. At other times, however, attempts to make an ape copy a design have failed.

In drawing on these observations, Rensch concludes that apes sometimes produce unordered scribbles. For a fraction of the designs and paintings, though, he admits that they have been influenced by factors which he does not hesitate to consider primitive esthetic sensations or sentiments, especially if one can judge from analogous works by human infants. He thinks that the physiological basis for sensations of elementary esthetic satisfaction is probably the same in higher animals and in man.

For animals as for human beings, guidance provided by the contours of a sheet of paper or by some preexisting lines eases the matter of composition and makes a contribution of positive emotive value. The contemplation of comfortably complex structures which are easily grasped is equally satisfying. A satisfaction in rhythms, symmetries, balanced compositions, and an adaptation to the format of a sheet or to preexisting traces is frequent in man; to a certain degree a similar satisfaction is detectable in higher animals.

In his book on the biology of art, Morris (1962) offers a number of observations on the paintings and designs of apes, and then discusses the evolutionary origin of human esthetic sensations. He recalls with some humor the impassioned and contradictory reactions provoked by the exposition of chimpanzee paintings organized in 1957 in London by the Institute of Modern Art. Since most of the English love animals but detest abstract art, they suffered a mental conflict on encountering these "works" by chimpanzees. Certain art critics spoke of an insult to human dignity, whereas others announced the birth of a new form of vital art. Only a few were fair in considering these paintings as unique documents, permitting a biological and evolutionary approach to the phenomenon of human art.

Morris thinks that man was first able to make drawings in soil and then on rocks, trees, and animal hides; starting when arms and hands were no longer used for arboreal locomotion, human beings were free to manufacture tools and to produce paintings on the walls of caves. Under experimental conditions apes can be induced to design and paint; why, then, have they not utilized these faculties under natural conditions? It is all the more astonishing, says Morris, because in some instances the apes have preferred painting to eating, and have been angered when they were prevented from painting. Painting is a useless activity which brings pleasure to an ape. This type of activity, one could argue, is one seen among animals who are not preoccupied by food and self-defense, as is the case for apes in the Zoological Garden. Under natural conditions only very young apes are still free of such vitally important preoccupations.

This line of argument is unsatisfactory for two reasons. Among birds and mammals, many playful activities exist among adults—not only among the young. Furthermore, one does not find even young apes making pictures in the dirt under natural conditions. Morris claims that higher apes have not developed their esthetic talents because they have neither necessity nor reason to do so. Man, in contrast, had a need for a means of perfecting his verbal communication and pictorial representation in order to organize the hunt, thereby increasing the chances of all participants against animals much stronger than they.

Here again Morris's arguments do not seem satisfactory. If man is the only animal to have developed an articulated language and some esthetic pictorial expressions, it is because the structure of his brain, notably of the cortex of the telencephalon, has given him that capacity. He has thus acquired as an outcome of his biological evolution the extraordinary and powerful means of cultural evolution. We do not think that it is possible to understand the evolution of man's pictorial art (as with other artistic expressions) without taking into consideration natural selection and biological evolution. Art is the property of our species: one of the new ways of expression arising from cultural evolution.

The First Manifestation of Artistic Expression in Man

Our knowledge of the first manifestations of artistic expressions of man is not ancient, and the number of described works is not high. Leroi-Gourhan

(1965) thinks that it was in about 1834, in the cave of Chaffaud in the department of Vienne, that Brouillet made the first discovery of a prehistoric work of art: an engraved bone showing two does. The celebrated cavern of Altamira was discovered in 1879 and that of Lascaux in 1940. Examples of paleolithic art have since been found in both Europe and Africa. In Asia only one such discovery has been made, near Lake Baikal.

The first engraved designs, nonfigurative or prefigurative, of the Chattelperonian period were produced about thirty-five thousand years ago. They consist of incisions on teeth and bones of reindeer. The admirable works of figurative art in the cave at Altamira date from the Solutrean period (about seventeen thousand years B.C.) to the Magdalenian (about ten thousand years B.C.). And the chief works of Lascaux also date from the ancient Magdalenian; that is, from about 15,000 B.C.

Certain of the paintings of Altamira and of Tuc d'Audoubert have an extraordinary power of expression. They were executed by exceptional artists. The esthetic sensation is so intense that they rank with the works of the best painters of recent centuries. It would be wrong, however, to say that art has been great from its birth, that artistic expression existed in man from the outset. Although the vestiges of prehistoric art discovered so far are not numerous, we are sure that the first artistic expressions appeared very late in the evolution of man. The first known works did not have the perfection of the best works at Lascaux. Furthermore, lest we forget, together with the marvelous deer, bison, and horses of the Magdalenian are many mediocre designs and paintings of the same epoch.

Before arriving at the production of paintings and statuettes, the skilled hands of prehistoric man permitted him to fashion, with increasing refinement, tools and flint or bone weapons. The question here concerns the expression of an esthetic sense. The prehistoric artisan sought to give agreeable forms to objects that he produced. Technical or functional progress does not explain the evolution in the manufacture of prehistoric tools and arms. Even today the artisan's pleasure with manual work is an essential motivation for most artists.

One finds a trace of that esthetic satisfaction among all who have a manual activity, whether useful or not, whether necessary or recreational. One may even say that a biologist who has performed a beautiful dissection is an artist.

Early man adopted the habit of decorating his tools, utensils, and arms.

Some bone spears from the Gravettian (twenty-two thousand years ago) carry engravings of horses and mammoths. The esthetic pleasure of the artisan's decoration of utilitarian objects and personal property was maintained nearly everywhere until the nineteenth century, and still exists today among populations which are not esthetically drained by industrialization—a process which has replaced manufacture in the true sense of that word. Throughout man's history the artisan's expression of esthetic sensations has been commoner than the production of works of art; the paintings at Lascaux were produced by a few exceptionally endowed individuals.

After Altamira and Lascaux, cultural evolution has permitted the refinement and diversification of artistic expressions not only by the invention of new supports, materials, and artistic tools but also by improved transmission of technical knowledge and the mass of visual acquisitions of the artist. Improved means of transportation, museum collections, reproductions, and television allows the modern artist—at least potentially—to become familiar with the art of the entire world and of all time. That is a direct outcome of man's cultural evolution. With respect to biological evolution, the cerebral capacities (otherwise knowledge) and the manual dexterity determined by the anatomical and histological structure of the brain (otherwise the technological means and the esthetic conscience and the whims of creation) have probably not changed much since the Magdalenian, and Lascaux. But it must be repeated: the chief works of Font de Gaume, of Altamira, and of Lascaux represent the result of a long esthetic evolution of man. These fine paintings do not represent a beginning; on the contrary, they represent a summit already attained.

Conclusions

In the *Encyclopedia* Diderot (1751) discussed the origin and the nature of beauty: "Everyone thinks about beauty: one admires it in the works of nature, one demands it in the productions of art. . . ." He called beauty "all that which contains within itself that which awakens in my senses the idea of communication. . . . Place beauty in the perception of communication and you have the history of progress from the birth of the world to today." But "there are probably no two men on earth who perceive exactly the same communication in the same object, and who judge it beautiful to the same degree. . . ." For Diderot, art which creates beauty is work for an intelligent cause, but biologists' specimen cabinets offer a large number of communications which are the results of fortuitous combinations: "Nature, in its play, imitates on a hundred occasions the productions of art. . . ."

When Diderot wrote that article for the *Encyclopedia*, a clear distinction was made between "brutes" (read "animals") and man. The ancient ideas of evolution had been forgotten or were not taken into consideration. The intent of encyclopedists, the essence of a marvelous century of enlightment, was to stimulate the faculties of well-informed persons with respect to knowledge, geometric and mathematical order, logical systems, and typological preferences. Diderot spoke of "that philosopher who was thrown by a tempest on the shore of an unknown island . . . (and who wrote) . . . upon seeing some geometric figures: 'Courage, my friends, here are human footprints.' "

More recently the great minds of modern evolution, first Lamarck and then Darwin and Wallace, have stimulated us to see and to comprehend the unity of the world, and to grasp our place and role in the world through the theory of biological evolution. To understand the unique place of man among all organisms, however, one must grasp the extraordinary importance and the incomparable power of cultural evolution. Cultural evolution has accelerated the pace of human evolution; during some tens of thousands of years the difference between *Homo sapiens* and other living organisms has grown so large that even evolutionists are hard pressed to explain them satisfactorily. One should not be surprised, then, that certain persons will not admit that evolution is able to explain the creative evolution of man and his culture; these persons postulate instead the existence of forces sublime, supreme, supernatural, unknown, and unknowable. At times they even resist efforts that are aimed at understanding and explaining human evolution.

We pretend neither to know nor to explain all the evolutionary steps that have resulted in modern man. We have surrendered (perhaps imprudently), however, to the temptation of applying our concepts of creative evolution to the expressions of man's esthetic sense. The power of creation manifests itself more than elsewhere in the cultural evolution of man—the supreme product of biological evolution.

Contrary to what Diderot (a man for whom we have great admiration) affirms, nature does not imitate art. Nor does art imitate nature. Man, however, forms part of nature. Man *is* a product of biological evolution, with sensory organs, muscles, and a brain that represent an improvement or enrichment by evolution of structures preexisting in other organisms. No other animal creates works of art. That does not mean, however, that art has fallen from the sky, or that in a twinkling it appeared in man. Art is not a novelty that arose ex nihilo, without a biological basis. Forms, rhythms, patterns, and colors have always existed. By evolutionary descent, long before the appearance of human artistic expressions, man acquired and then transmitted the genetic capacity for esthetic perception.

Esthetic perceptions, as we have seen, play a large role among numerous organisms, including man's relatives. In nearly all artistic expressions, the fundamental influence of the forms, rhythms, designs, and colors of our natural environment can be seen. Human art, however, transcends the esthetic sensations of animals and the esthetic expressions of bower birds and higher apes.

The art of man is unique: Nothing comparable to it exists in any other animal. It did not always exist, however, even in man. The latter claim raises questions: From what time can we speak of "man?" Where and at what stage can a line be drawn between "ancestors" and "man?" For an evolutionist, these are badly posed questions. It is impossible, if one truly accepts the evolutionary principle of *panta rhei*, to designate a precise moment at which man appeared. From our evolutionary point of view, we refuse to designate a border (in what would be an unrealistic typological move) between *Homo sapiens* and other organisms. But we can study those elements which characterize the evolutionary path to modern man. We think that art is both a property of and a new capacity of man. In the preface to the first volume of *Univers des Formes*, Malraux (1960) said that one tied art first to imitation and idealization, then to spiritualization—Christian idealization. He thinks "that all idealization tends to create admiration and relates to life, even immortality . . . that the stylization of high epochs tends to create veneration and relates to eternity . . . a pointed geometry seems to symbolize the victory of spirit over chaos." For Malraux, "idols having ceased being effigies, the problem that their creation poses to the artist is clear: to create a figure which is other than that which it represents. Just as we cannot admire as art a picture which is merely realistic, a Sumerian was not able to pray to a statue in which he saw only the image of a man." For Malraux, the god Abu or the goddess Tell Asmar of the third millennium (now found in the museum of Baghdad) does not suggest a real person; for example, the enormous eyes are not an artist's blunder. Artistic discrepancies are a means of creation. The Sumerian sculptor elaborated the head in order to create an idol for a temple: if he simplified the forms he created, it was to release man from his humanity.

Man is the only organism who knows of his evolutionary origins and who has developed (largely through communication, notably by his language) a conception of the perpetuity of his species. Pictorial art, sculpture, and other artistic expressions represent a means of communicating with the entire species, even with persons still unborn. If esthetic sentiments and esthetic expressions exist among animals, and if human art has (as it seems) an evolutionary origin, man remains alone, nevertheless, in having created a means of communication that bears witness to an awareness of our origin and to the perpetuity of our species.

Apes may create paintings that have an esthetic attraction, but they remain incapable of imitating objects or designs. To do so requires an awareness man alone has acquired by cultural evolution during the course of humanization.

Can man's artistic activities be explained as an animal behavior, in light of what is known of biological evolution? For Morris (1962) the answer is rather simple: as soon as man developed a true language, a device that was needed to organize hunts and to describe objects and sentiments, the door was open for the pictorial representation of these objects. That was when prehistoric art appeared. It was used for the description and the planning of the hunt, just as magic was used for ensuring success. Still, one notes, especially at Lascaux, the esthetic element. Today, only the esthetic motivation remains: therein lies the success of nonfigurative art.

Dissanayake (1974) postulated that the artistic comportment of man should have, as do most biological functions, a selective advantage, because this behavior occurs in nearly all human societies. Art is an activity of play that utilizes energy that is not demanded by other, essential activities. Like play itself, artistic activity is neither a necessity nor is it functional, nor has it a use: it produces pleasure. Instead of seeing the relations between play and art ontogenetically, Dissanayake takes into consideration their evolutionary phylogenetic relationships. Early man's impulse to play provided the basis on which artistic behavior was developed. Later, that artistic behavior was reinforced by a sense of order, novelty, and rhythm, even as the social and cultural meanings of artistic works were reinforced. Because art reinforces social values, a selective pressure worked in its favor. All behavior that reinforces sociality has a selective value and, consequently, is maintained.

Another factor has been important in the development of artistic activities: pride of workmanship. The discovery of prehistoric shops in which tools and arms were made suggests that certain individuals specialized as tool makers. They developed superior manufacturing techniques and passed these on to their children. For amusement, some of these artisans decorated tools as early as the Chattelperonian (35,000 B.C.) Consequently, we know that man had even then an esthetic sense, and an appreciation of decorated tools and arms. The fashion rapidly spread. In the Gravettian (22,000 B.C.) one recognizes an Isturitz spear with the engraving of a horse. Parietal art probably developed from decorative objects. From the Solutrean (17,000 B.C.), we find the beautiful and vigorous paintings of Lascaux and Altamira. At the mid-Magdalenian (12,000 B.C.) appears the astonishing clay bison of the Tuc d'Audoubert. In La Madeleine, carvings on a reindeer horn represented a magnificent bison with its head reversed.

Most authors admit that parietal art and sculptures have a role in magic. That role does not diminish the esthetic sense of artists and of those who considered artistic works to be important. Art at the beginning of the historic period—for example, that of Sumeria of 3000 B.C.—exhibits a refinement of technique, but no increase in the power of esthetic expression. The same remains true today. Much modern art has a smaller force of expression, even of esthetic value, than the paintings of Lascaux.

Biologically, the conditions for the creation of esthetic works were present in the Solutrean. Cultural evolution has since furnished a diversification of supports, materials, tools, and techniques. There has also been an extraordinary increase in the number of persons who are able to contemplate artistic works.

Art has always been a means of communication. Its role today is more important than ever before. Superspecialization, the cold culture of concrete, the monstrous agglomerations of millions of individuals, and the brutal acts seen in the media create grave dangers for humanity. Faced with cultural degradation, art, by becoming even more than before a means of humanistic communication, can mitigate the alienation and the disequilibrium of modern man.

The above postulate demands participation. One cannot prescribe even the role of art. We cannot erect a code of beauty; we cannot erect barriers for the artist's imagination. We can, however, offer some fundamental principles based on our biological and cultural evolution. Humanistic art is art that enhances the growth of those tendencies that have been revealed during the evolution of man. Three points may be emphasized:

First, biological evolution produces in sexually reproducing diploid organisms an enormous heterogeneity. Each individual has its unique genotype, one that differs from those of all others. Among human beings, and them alone, cultural evolution has created conditions of life which permit a growing differentiation of attitudes, functions, and activities for each individual. Each individual should be permitted to fulfill his life in harmony with his genotype, to live to the optimum permitted by his genetic endowment. Humanistic art should promote and sustain an enriched cultural liberation. All tendencies toward uniformization (already feared by many art critics) should be condemned. Such a tendency is seen, for example, at the Bienalle of Venice, where the different paintings and sculptures vary from one another only slightly. That the galleries of Tokyo, Milan, Paris, London, and New York should promote same types of painting is a dangerous antievolutionary and antihumanistic aberration.

Second, a total human diversification could result in chaos. For variety not to oppose humanism, the other fundamental tendency of human evolution must be reinforced: the development of harmonious social structures. The liberation and development of individual faculties are not possible without a price. To ensure an equilibrium between the constraints necessary for respect for the liberty of others and the optimal and free realization of the potentiality of each individual genotype is difficult. Art—the different arts— is the means of communication and social contact, which are increasingly necessary for the humanistic evolution of social structure. Forms of art leading to aggression and antisocial behavior bode ill.

Third, man and his art have resulted from the interplay of biological and cultural evolution. Hence, one can claim that nature and other organisms provide certain keys for the judgment of natural and pleasing esthetic harmonies. One must not lose contact with nature; it must not be destroyed. The evolutionist demands the maintenance and intensification of those esthetic harmonies of forms, colors, rhythms, and designs which nature developed long before the arrival of man. One of the great humanistic roles of the artist is to make natural beauty visible to those who either cannot see it or can see it only poorly, because beauty is a creation of evolution.

The arts should prove to be more and more indispensable in combating the ugly, aggressive, and antisocial forces that tend to destroy the social cohesion of our species. Cultural evolution has led humanity to the edge of self-destruction. Only by a cultural evolution even more advanced than ours now is can we avoid catastrophe. The arts have an essential and new function in this progressive evolution: art is a product of cultural evolution; it must be a factor in its own future development.

CHAPTER 7
Human Equality and Genetic Diversity

The ideals of *liberté, égalité,* and *fraternité* are widely known, even if they are not always realized in the modern world. The intellectual antecedents of these ideals lie in classical Greece and in the teachings of Christ, but it was in eighteenth-century .France that they were formulated as political principles. Thomas Jefferson, who was an intellectual descendant of the French Enlightenment, wrote in the Declaration of Independence of the United States in 1776: "We hold these truths to be self-evident, that all men are created equal. . . ." A corresponding phrase in the Declaration of the Rights of Man, promulgated by the French National Assembly in 1789, is: "Men are born and die free and equal in rights. . . ."

Yet it is often claimed, even by reputable scientists, that biology has demonstrated that persons are born unequal. This is sheer confusion. Biology has shown that persons are genetically different, not that they are unequal in their rights. Genetic diversity and human equality belong to different realms of ideas. Human populations, like populations of all sexually reproducing organisms, are polymorphic for many genes. Therefore, the probability that any two individuals have the same genotype by chance is practically zero.

Only monozygotic twins, which arise by an asexual duplication of a sexually produced zygote, are usually identical in their genetic endowments. Human equality, equality of opportunity, and equality of rights, however, are not predicated on genetic identity. By the same token, human inequality does not follow from genetic diversity. Identical twins may engage in different occupations and achieve different socioeconomic status. Genetic diversity is a fact of nature. Equality and inequality are not biological phenomena but, rather, sociological designs; they stem not from genes but from ethical, political, or religious wisdom or unwisdom. Equality may be granted to all humans, or only to citizens of some countries, or members of some social classes, or those favored by a monarch or dictator. Genetic diversity cannot be brushed away even if this were desirable (which it is not); equality, on the contrary, can be lost by the stroke of a pen.

The Myths of Environmentalism and Hereditarianism

Genetic diversity on the one hand and social equality or inequality on the other are, in principle, independent. They are, nevertheless, not mutually irrelevant. Any society faces the problem of persuading its members to perform different kinds of socially necessary or useful work. The problem becomes more and more complex as simple societies become technologically more advanced. The variety of occupations and professions is greater in the latter than in the former. The rationale of the doctrine of human equality is not to make everybody alike and engaged in the same occupation. Quite the opposite: this doctrine permits recognition in practice of the diversity of individual tastes, preferences, and abilities. Equality would be superfluous if all human beings had the same tastes and abilities. Persons could then be assigned to different occupations by a lottery or any correspondingly arbitrary method. Because persons are not alike, the doctrine of equality warrants a recognition in practice of the diversity of individual tastes, preferences, and abilities. All persons should be permitted to aspire to and work at the occupations they prefer and induced to develop their individual abilities and talents. It is because people are not all alike that the denial of equality leads to a waste of human talent, as well as to pervasive personal discontent.

Two explanations of human diversity had been advanced even before there was a science called biology. One ascribed human differences to environments, the other to heredity. These hypothetical explanations still have their adherents. In their extreme forms they can be labeled the myth of tabula rasa and the myth of genetic predestination. In its modern form, the tabula rasa myth has a strong appeal to many social scientists. All human beings are deemed to possess the same potentialities at birth (or at fertilization); they become different owing to upbringing and training, to psychological conditioning by varied circumstances of their lives. The conclusion that logically follows, and that is actually defended by the behaviorist school of psychology, is that individual human beings are interchangeable. When brought up in a certain way, any individual could be prepared for any function or role in the social system.

The myth of genetic predestination, in its modern version, appeals to racists and to a minority of biologists. They assume that to adequately perform the duties of a given social position, one must be born with genes appropriate for that position. Education and training are largely powerless to change that which genes have decreed. The socioeconomic structure of a society is, then, a reflection of its genetic composition. The rich and the poor, the powerful and the humble, the advanced and the backward are what they are because their genes have decreed it so.

The caste system in India was implicitly based on a prescientific idea of genetic predestination. Bose (1951), who is not a defender of the caste system, explained it as follows:

"The careful way in which the tradition of close correspondence between caste and occupation was built up is a clear indication of what the leaders of Hindu society had in mind. They believed in the hereditary transmissibility of character, and thought it best to fix a man's occupation, as well as his status in life, by means of the family in which he had been born."

One can say that the caste system in India was the greatest genetic experiment ever performed with human beings. For more than two millennia a society endeavored to induce genetic specialization in caste populations. Indeed, each caste (or rather subcaste) not only had a different social status but engaged by force of tradition in a different occupation or profession. It now appears that the grand genetic experiment was a failure. Although modern India has far to go in abolishing caste inequalities, all castes have already produced persons competent in nontraditional training and in nontraditional occupations.

The Genetic Repertoire

We must now explain an elementary concept of genetics—one which is curiously difficult for persons to keep in mind and to think of properly. This concept is that genes do not each determine one characteristic of the adult organism, independently of the environment and of the other genes of the same organism. To be sure, at the biochemical level, each structural gene codes for one and only one polypeptide chain. But at the level of the phenotype, of the morphological and physiological manifestations which we are discussing, genetic effects often amount to a conditioning rather than a predetermination. This is because between the genes and their phenotypic effects, processes of growth and development intervene. In these processes, gene products interact with environmental influences and with the products of other genes. If one could have an absolutely constant environment and a single genotype absolutely uniform except for one pair of alleles, then one could say that *under the conditions of the experiment* each of those alleles determines a certain characteristic or characteristics. In reality, such constancy is not attainable. Therefore, a gene or genes, interacting with a range of environments and of other genes, makes possible a certain potential genetic repertoire, a certain norm of reaction, a certain range of phenotypic manifestations.

The genetic repertoire, or norm of reaction, is not completely known for any genotype. This is so because to know it completely one would have to observe the phenotype (in its entirety) produced by the genotype in all possible environments. This is impossible because the variety of environments is infinite. Still, it is of great importance to know as much as possible about the norms of reaction of human genotypes, and of those of agricultural plants and animals. Experimental medicine, hygiene, pedagogy, and experimental agronomy all study norms of reaction. In principle, the manifestation of any human genotype is manageable by environmental modification. In practice, this is not always possible. The manifestation of a human genetic disease, phenylketonuria, can be largely cured by placing the affected child on a special diet containing as little of the amino acid phenylalanine as possible. What is "cured," of course, is the phenotypic manifestation; the gene responsible for the defect remains unaltered. Such "cures" are being discovered for more and more genetic diseases, but for many no "cures" are as yet known.

Genetic predestination is a myth because the phenotypic manifestation of human genotypes is a function of human environments. What is good and superior in one environment may be bad and inferior in another, and vice versa. Certainly we do not know, and may not know for a long time, those environments that are specifically favorable for certain genotypes. Racism is, however, evil and foolish because it assumes, without any basis in fact, that the genotypes of certain human *populations* are intrinsically and irreparably inferior to those of other *populations*. The tabula rasa theory is also a myth because it fails to recognize a fact mentioned repeatedly throughout this book: in mankind, as in any sexually reproducing species, no two individuals are genetically identical. Human populations are polymorphic for all kinds of genetic characters: biochemical, physiological, morphological, and psychological. This polymorphism brings about an enormous array of norms of reaction. Associated with a variety of environments, these repertoires result in a virtual infinity of manifestations. Some grossly pathological genetic variants excepted, these manifestations give mankind the esthetic and spiritual luxuriance without which the world would be unbearably tedious and boring.

IQ and the Heritability Controversy

Although neither author of this book is a psychologist, we are obliged to give attention to the genetic aspects of variation in the so-called intelligence quotient, or IQ. This is because the recent work of some psychologists, particularly Eysenck (1971) and Jensen (1969, 1973), have provoked a storm of controversy which is of undoubted interest to students of human evolution.

The validity of the IQ as a measure of intelligence is itself in dispute. IQ tests, first proposed by Binet and Simon in France in 1905, have been developed further by many psychologists in many countries. An enormous number of individuals, particularly children, have been tested. Most authorities concede that an IQ test measures not a physiologically or genetically unitary mental ability but, instead, a composite of an unknown number of traits, some tightly and others only loosely correlated with each other. It is also conceded that none of the many IQ tests so far invented can give meaningful results, except in the particular cultural and linguistic group for which it was devised and in which it has been standardized. So-called culture free tests have been proposed, but none of them has thus far proven valid. Why, then, is IQ testing still being practiced? The justification is a pragmatic one: IQs provide moderately good forecasts of the success of children and young persons in schooling, but only within a common educational system prevailing within a given country. It is not impossible that those who are "intelligent" in French lycées would not be considered so in schools in mainland China, and vice versa.

That the differences in IQ among individuals of the same population are both genetically and environmentally conditioned is established beyond doubt. Whether differences in the average IQs of different populations (social classes, castes, or races) are also in part genetic (and if so, what portion of the observed differences is due to genes) is quite another matter, these two sets of differences must not be confused: variation between individuals *within* a relatively homogeneous population on the one hand, and the average differences *between* populations on the other. Much of the controversy surrounding the issue of genetic components of the variation in IQ results from a confusion of these two very different sorts of variation.

Physical resemblance between identical twins (monozygotic twins), which sometimes amounts to near identity, has been known for a long time. The resemblances between fraternal twins (dizygotic twins) is much less striking. The same is true of the IQs of identical and fraternal twins. The results of many independent studies on IQs of twins and other relatives have been summarized by Erlenmeyer-Kimling and Jarvik (1963). The median correlation coefficient between IQs of identical twins brought up in the same family may be as high as +0.87; for fraternal twins of the same family the correlation is significantly lower, +0.56. Monozygotic twins have identical or at least extremely similar genotypes; the genotypes of fraternal twins are, on the average, as different (or as similar) as those of brothers and sisters who are not twins. The greater similarity of identical twins is supposedly caused by the similarity of their genes. Still, identical twins are not identical in IQ; +0.87 is not +1.00, a perfect correlation. The differences between the IQs of identical twins are environmentally caused.

The above conclusions have been challenged on the grounds that the environments of identical twins are more similar than those of fraternal twins. Parents, teachers, and everybody else treat identical twins more similarly than they treat fraternal ones. This objection is invalid. Indeed, parents and others do not know for certain whether a given pair of twins is monozygotic or dizygotic. If they treat identical twins more similarly, this is because twins of this sort are so similar in appearance and behavior. In other words, the genetic constitution of a person is in part responsible for how that person is treated by others. Significant data have come from monozygotic twins who were separated in early infancy and subsequently reared in different families.

The IQs of identical twins reared apart show a correlation of +0.75. This is less than for the same sort of twins reared together (+0.87), but still higher than for fraternal twins reared together (+0.56).

Important data come also from correlation studies of the IQs of relatives who are not twins—parents and children, brothers and/or sisters reared together or reared apart, cousins, and still others. Especially significant are studies on foster children, adopted by persons who are not their biological relatives. The IQ of foster children are more strongly correlated with those of their biological parents than with those of their foster parents or with those of the biological children of the latter. These findings can hardly be explained in any other way than that, within the range of the environments represented in the materials studied, the genetic determinants of IQ are more powerful than the environmental ones. Of course, this range of environments is a limited one. The adopted children and their biological and foster parents are living in the same country and often belong to the same social class. We do not know what would happen if children of middle-class Europeans were adopted by Eskimos, or by Tibetans, or by a tribe of primitive hunters—or vice versa. It could well be that the foster children in such cases would resemble their foster parents more than their biological ones.

Attempts have been made to derive from the above data quantitative estimates of the genetic component of variance among IQ scores, and of the environmental component. Such estimates are known as estimates of heritability. Jensen (1969, 1973) reviewed all available data and obtained a heritability estimate of 81 percent. That is to say, 81 percent of the variation in IQ is due to genetic differences between individuals and only 19 percent is the result of the different environments in which these individuals developed. This estimate has been seized on by racists and other believers of the heredity myth. Persons are intelligent or stupid because of their heredity; the environment, including education, they say, has little or no power to modify inborn quality!

The meaning of heritability estimates is often misunderstood. Heritability is not an intrinsic property of IQ or of any other variable character. It is a property of the population. It is subject to change when the population becomes more or less genetically variable, or comes to live in more or less diversified environments. Imagine that the genetic basis of IQ is uniform in all human beings; the IQ phenotype will still be variable because of environmental variations. Heritability in this case will be zero. Suppose, however, that genetic variability exists but the environment is exactly the same for everyone. Heritability in this case will be 100 percent. If the environment is different for different persons, heritability tends to decrease. Finally, suppose that heritability in a certain range of environments is 100 percent; this would not mean that IQ cannot be changed by environmental means. New environments—for example, novel educational techniques—may alter the IQ phenotype.

Scarr-Salapatek (1971a, 1971b) has obtained data suggesting that the heritability of IQ is higher in whites than in blacks, as represented in Philadelphia schools. Families of schoolchildren were classified according to their socioeconomic status. Black families were more often of low status than were white families. A greater proportion of the IQ variance in white families than in black ones was attributable to genetic causes. This is interpreted to mean that genetic differences in IQ manifest themselves in the phenotypes of persons who develop in more favorable and stimulating environments but remain unexpressed under conditions of poverty.

Jencks and his collaborators (Jencks, 1973) have analyzed the high estimates of IQ heritability obtained by Jensen into their component parts. Environments, particularly the cultural environments of human beings, are not independent of genotypes. People are able, within limits, to choose or create their environments. This is genotype-environment covariance. The IQ phenotype of a child is influenced by the genotype of the child and also by the home environment provided by its parents. Such environments are more favorable and stimulating in prosperous than in poor homes, and also more favorable if the IQs of the parents are high than if they are low. The IQs of the parents are due to their genotypes as well as to the environments in which they grew up. Consequently, there is a network of covariances, of paths interconnecting parental genotypes with parental phenotypes, home environments, socioeconomic status, genotypes of children, and finally, phenotypes of children. If the effects of other variables are identified and removed, it appears that the diversity of genotypes among children accounts for only 40 percent of the phenotypic variance of their IQs.

Group Variability

It is, in our opinion, established beyond reasonable doubt that the diversity of IQs among individuals of the same socioeconomic status and in the same genetic population has a far from negligible genetic component. This is true regardless of what estimate of heritability—80 percent or 40 percent—one chooses to take. The hottest polemics, at least in the United States and England, have arisen not in connection with individual differences but with the average differences between groups of persons—social classes and races. It has been shown repeatedly that IQ averages are higher in prosperous socioeconomic classes than in those that are poverty stricken. Numerous studies have shown that the average IQ of blacks is ten to fifteen points below that of whites in the United States.

Because the heritability of IQ seems to be high, some persons have hastened to conclude that group and race differences in IQ are of genetic origin and independent of the environments in which these groups or races live. This conclusion is a non sequitur. The heritability of individual variations in IQ (or in any other variable trait) is valid only for the range of environmental variations found in the population. The work of Jencks, cited above, has shown that even with this restriction, the causation of the IQ phenotype is complex. The heritability of individual variations does not necessarily tell us *anything* about the genetic basis of differences between groups which live in different environments. This is particularly true for the white and black populations of the United States. The environments of these populations differ in obvious ways: average income, housing, and schooling. More subtle, but perhaps no less important, differences are the result of cultural factors. These, of course, reflect the consequences of slavery in the past and a long history of social discrimination and popular prejudices which have not yet disappeared.

To what extent, if any, genetic factors are responsible for race and class differences in average IQ is, we believe, an entirely open question. In theory, the problem could be resolved only if the physical and cultural environments of all persons were to become completely alike. Perhaps less unrealistic is the possibility that average environments might become similar for all social classes and races, while the range of environmental variation remains as it is now for middle-class Europeans and Americans. Jensen (1973) made an unsuccessful attempt to pin down the environmental factors responsible for average IQ differences between black and white Americans. In our opinion, a failure to identify such factors is not evidence that intergroup differences are therefore genetic.

Equality of Opportunity and Meritocracy

Persons are born equal in rights; they are not born genetically identical. Various degrees of equality and inequality of opportunity exist in different human societies. Equality and inequality are sociological, not biological, phenomena. Nevertheless, sociological arrangements have genetic consequences. Many sociologists, and virtually all politicians, are unaware of these consequences.

Under equality of opportunity, the role that a person plays in society is achieved through ability, combined with a willingness to exercise this ability in an occupation. With inequality of opportunity, roles are ascribed rather than achieved. The ascription occurs on the basis of race, social class, and family connections. Equality of opportunity leads to meritocracy. The sociologist Eckland (1967) defines meritocracy as "a society in which positions are allocated on the basis of talent, plus effort, rather than class or social advantage."

Traditional castes in India assigned individuals to various roles on the basis of birth. A Brahman was a Brahman because he was born a Brahman, not because his individual abilities qualified him to be one. Someone born in a low caste could not become a Brahman regardless of his abilities. Social mobility and gene exchange between castes were illegal. Traditional closed-class societies of feudal Europe also discouraged social mobility and intermarriage between classes, but the rigidity of these customs varied from place to place and from time to time. Gene exchange, intra- and extra-marital, took place. The historical trend, at least since the Renaissance, first in Europe and eventually throughout the world, has been from closed- to open-class societies. With the growth of capitalism, the accumulation and subsequent loss of wealth became powerful forces of social mobility and, consequently, of gene flow between social classes. Equality of opportunity and meritocracy are now accepted in words, though often not in deeds, as desirable social arrangements in developed countries.

The fatal flaw of caste and closed-class arrangements is that they lead to a waste of human talent. A society benefits from the fullest possible realization of genetically conditioned, socially useful abilities of its members. If Beethoven had been born into a poor peasant family, he might have played music at village festivities, but we might not have had his symphonies. If Darwin had been a sailor or a messboy instead of the naturalist on the

Beagle, he might not have written the *Origin of Species*. How many talents of diverse kinds are even now frustrated in the slums of our large cities?

We do not make the naive assumption that there exists a special gene for every "talent." The genetic basis of personality and of talent is an emergent product of the entire gene constellation. The realization of the gene constellation in the phenotype occurs through interactions with the environments which a person meets during his lifetime. Without in any way underestimating the importance of the environment, one may nevertheless say that Mozart, Beethoven, Darwin, Einstein, Racine, and Anatole France carried gene constellations the likes of which are not common in any human population.

Genetically conditioned talents of all kinds are dispersed throughout the social structure, in all strata of the social pyramid. Without equality of opportunity many (probably most) of these talents remain unutilized. This would be true even if the incidence of some abilities and talents were greater in some social classes than in others. The evidence for this claim comes, among other sources, from the destruction of old elites during revolutionary upheavals, as in France and, more recently, in Russia and China. What was feared in each case to be the intellectual decapitation of a nation did not result in any permanent genetic impoverishment. New elites were promptly recruited from the formerly suppressed masses.

Equality of opportunity offers the only known way in which the pool of genetic talents and abilities present in human populations can be fully utilized. Any person should be entitled to aspire to any position or role in society. Obviously, not everyone will realize his aspirations. This will depend, among other things, on his genetic endowment. Ideally, every person should elect the career for which that person is best qualified. For a tone-deaf person to aspire to a career as a musician is silly, as it is for a frail boy to aspire to be a professional boxer. Equality of opportunity would, however, maximize the probability that individuals with pronounced abilities of different kinds will become located in their proper niches in the social edifice.

Because equality of opportunity was denied for so long to so many persons, it has become the watchword of those who champion justice and human dignity. Equality of opportunity leads, however, to meritocracy: it does not by itself lead to equality of status. The prestige that is attached to some occupations is greater than that attached to others, and economic rewards are very unequal. Power and influence often accompany prestige and money.

The number of vacancies in the most prestigious and rewarding occupations is limited. Hence, a competition, often a brutal competition, develops for the occupation of these vacancies and achievement of the status accompanying it. Yet to be "equal" in fact, as well as in name, persons feel the need of both equal opportunity and equal status.

Meritocracy and Aptitude Aggregations

Biological evolution is governed by biological laws; the evolution of human societies is governed by sociological laws. To think, however, that biology and sociology are independent would be a grave mistake. They are, as we saw in chapter 1 and elsewhere, interdependent. They are parts of the same cybernetic system. We have pointed out above that the denial of equality of opportunity in caste and traditional closed-class societies leads to a waste of human talents, many of which may be genetically conditioned.

Partisans of class societies claimed that the higher classes concentrated genetic talents which were rare or absent in the lower classes. This is, however, precisely what traditional closed classes or castes cannot achieve. The reason is simple: Mendelian segregation. Suppose, for the sake of argument, that some sort of aristocracy has been originally formed by the selection of individuals with some sort of talent. However, neither individual human beings nor populations of human beings are homozygous for all their genes. Generation after generation, there will appear within the aristocratic class individuals with characteristics unlike the class's founders. If these individuals are removed to other classes—in other words, if selection continues in all generations—it might be possible for the concentration of talents to be preserved, or even increased. A scion of an aristocratic family "inherits" the aristocratic social status, often regardless of whether he has inherited the intellectual talents of his ancestors. More than that, natural selection after a class or caste has been formed may favor characteristics other than (and possibly quite the opposite of) the original ones. What follows is a rapid erosion of the elite, or specialized, genetic endowments with which the class was formed. This erosion was recognized before any genetic understanding was possible; it was referred to as "degeneration" or "decadence." "Three generations from shirt sleeves to shirt sleeves" conveys a similar understanding.

The relaxation of caste rules and the transition from closed to open class structures increase equality of opportunity and lead to meritocracy. Let it be made clear that, although equality of opportunity has grown to various extents in various countries, it has not become complete anywhere in the world. The work of Jencks and his collaborators, already mentioned above, has analyzed the genotype-environment covariances which impose limits on equality. For example, they find that in the United States one of the ways economically successful families try to help their children retain their privileges is by making sure that their children get a good education. Such efforts are moderately successful. The correlation between a white child's educational attainment and his father's occupational status is almost 0.50.

No doubt this correlation is even greater in many countries other than the United States. Different kinds of education were traditionally designed for children of different socioeconomic classes, or else education was a privilege of the upper class alone. In a meritocracy, an upper-class child has a double advantage: he may inherit not only superior genetic abilities but also a family environment propitious for the realization of those abilities.

Several authors, especially Scarr-Salapatek (1971b), have pointed out that under meritocracy with only partial equality of opportunity, genetic differences between social classes are likely to be greater, not smaller, than between traditional closed classes. In the words of Scarr-Salapatek, "The greater the environmental equality, the greater the hereditary differences between levels of social structure. The thesis of egalitarianism surely leads to its antithesis in a way that Karl Marx never anticipated."

It seems that we have arrived at a discouraging conclusion: that equality of opportunity merely replaces social classes based on accidents of birth by genetically differentiated (and thus even more unalterably fixed) social classes!

While the above conclusion is probably inevitable given the present social structure of capitalist (and, it would seem, also of Soviet-type communist) societies, one may speculate that it would not be inevitable in a society that had reached a state of complete equality of opportunity and social status (see Dobzhansky 1973b). Under complete equality there will be formed aptitude aggregations, which in important respects will be novel phenomena, unlike the existing social classes. Equal opportunity should create a situation where every trade, occupation, and profession will at any given time include all or most of the persons who are genetically qualified for it. None of the aptitude aggregations will, however, consist of individuals homozygous for all the

genes that make these individuals prefer their respective occupations. Therefore, in every generation Mendelian segregation and recombination will produce individuals whose genotypes favor occupations other than those of their
parents. In traditional social classes, even the relatively open ones, parents
endeavored to make their children "inherit" their occupations and their sta-
tuses. The aptitude aggregations should exchange parts of their progenies in
accord with the aptitudes and proclivities of the latter.

Human Diversity: A Blessing or a Curse?

Every individual has a unique genotype. His genotype, and the environment in which he develops, makes the individual able to perform well in
different kinds of work. Some socially valued abilities are, however, relatively abundant, while others are scarce in human populations. Rather few
persons, for example, can become musical virtuosi or conductors of sym-
phony orchestras. Many more can perform well, or at least acceptably, as
farmers, manual workers, or clerks. In a meritocracy, the possessors of abilities that are rare but socially valued are likely to achieve high status and
material rewards. High ability tends to be concentrated in those positions
that bring greater prestige, power, and larger incomes. This is even truer in
societies that permit social mobility than in rigid class societies. Paradoxi-
cally, equality of opportunity not only does not lead automatically to equality of status, but may, in societies built on meritocratic principles, lead to
exaggerated and even more rigid status inequalities.

It would be an oversimplification to rank human abilities on any single
scale. Is an artist superior to a scientist or to an athlete, or vice versa? Is an
individual with a higher IQ always more valuable to society than is his
neighbor who has a somewhat lower score? Persons may achieve excellence
in artistic, intellectual, athletic, administrative, commercial, and other ca-
reers. An individual dedicated to music or poetry may be quite uninterested
in athletics. Nevertheless, prestige and material rewards may be achieved in
any one of these and in other occupations. This is not contradicted by a
statistical predictability of occupational and economic success from measurements of IQ and scholastic ability. It is not necessarily true that a moderately low IQ (not, of course, mental retardation) is a serious impediment

for high achievement in, for example, professional sports, commercial activities, or politics.

Why should it be that almost anyone can be trained as a manual laborer, but only a few can learn to conduct symphonies? Mankind is both a creator and a creature of culture; human beings are genetically adapted to live in environments brought into being by culture. The basic adaptive trait of a human being is the ability to be trained and educated for whatever work or occupation a society needs. Persons can be trained for many more diverse kinds of work than any animal, wild or domestic, can possibly be. Educability and trainability have been developed by the unrelenting pressure of natural selection in the course of human evolution. And yet educability is not limitless. Some individuals have special abilities that other individuals simply do not have, or have abilities greatly exceeding the population's average. Educability is, however, supplemented by genetic diversity. Some individuals take to certain kinds of training and work with ease and pleasure, while others have little or no success in the same kinds of training. The myths of tabula rasa and of genetic predestination are both deceptive; the truth is found in between.

We have pointed out that genetic diversity is a biological phenomenon, while human equality is a social rule based on ethical and religious principles. A human society gives to or withholds from its members equality of opportunity and equality of status. Different human societies, present and past, have adopted a variety of solutions for problems of equality, ranging from caste and rigid class structure to various degrees of equality of opportunity and meritocracy. In technologically developed societies the equality of status is, so far, a most difficult, unsolved problem. It is, of course, largely an ethical and political, and only marginally a biological or genetic, problem.

We can grant that different individuals have different abilities, and that such differences are in part (only in part!) genetically conditioned. But, in the words of Scarr-Salapatek (1971a), these differences "can simply be accepted as differences and not as deficits. If there are alternate ways of being successful within a society, then differences can be valued variations on the human theme regardless of their environmental or genetic origin." Anyway, these differences do not dictate the degrees of status and the economic inequalities now found in all technologically developed societies. Manual labor is not intrinsically inferior to intellectual labor, even though more persons can be trained for the former than for the latter. Efforts to uncover rare abilities need not detract from appreciation of the more widespread ones. This may be hard to accept for those who were brought up in class or meritocratic societies. We feel, however, that this view is desirable on ethical grounds. Moreover, political action in recent history seems to be moving toward not equality of opportunity alone but toward economic and status equality as well.

CHAPTER 8
Conclusion

All persons, except for some diehards, believe that the world is round; the few exceptions are to be found in the Flat Earth Society. Similarly, most educated persons (and this includes high school education) believe that human beings and other living organisms have evolved from common ancestors; the more similar the present-day forms, the more recently have they shared a common ancestor in the past. Because all forms of life utilize nucleic acids for the storage and retrieval of genetic information, and because the code by which this information is translated into protein structure is universal, it is reasonable to believe that all life on Earth had a single ultimate origin. This belief should be tempered: The ancestors of present-day life may have been exceedingly competitive, eliminating other, less efficient early forms.

Some persons, in a manner similar to the members of the Flat Earth Society, resist the notion that life has evolved. The bulk of these persons do so on purely religious grounds: to them the Bible represents God's word. It is a book to be interpreted literally, a book of authority whose contents must be accepted on faith. Except for the missionary zeal exhibited by these fundamentalist antievolutionists, they might best be left unmolested. Life, everyone will admit, is complex and stressful; for those who can retain their sanity only by believing that a Supreme Being has matters well in hand, let them believe. Should anyone, for the same reason, wish to believe that all cats are female and all dogs are male—by all means, let them believe, too. Let them believe, and let us wish them well.

Our objection to antievolutionary views arises when persons attempt to inject their religious views and their articles of faith into science education.

Science, as it has developed in the Western world, is a collection of hypotheses that have not yet been disproved; each of these hypotheses, however, is stated in a manner that allows for its own rejection: each hypothesis is testable. Hypotheses that are not testable—those that can account for any possible observation—do not belong in the realm of science. And for that reason, an all-powerful God who can perform any conceivable act and can work any imaginable miracle must be excluded from science education.

Not all opponents to evolution are fundamentalists; some are merely opposed to *Darwinian* evolution, to the notion that evolution has occurred through gradual changes in living beings—changes brought about by the slow remodeling of the gene pools of isolated local populations. We have attempted to show in earlier chapters that Darwinian evolution is unavoidable and that alternative suggestions invariably resort at some point (either explicitly or implicitly) to unproven inner forces or to vague external ones that in unexplained ways shape each species' future. The exception to this charge is the neutralist hypothesis of Kimura, which merely says that a great deal (perhaps most) of molecular change is a matter of indifference for the life, death, reproduction, or sterility of individuals. This hypothesis, which is a scientific hypothesis, has made predictions many of which seem to be supported by observations. Critically decisive data have not yet been collected regarding the neutralist hypothesis, however; for example, many genetic changes (such as chromosomal inversions) which are known to be nonneutral behave in the aggregate and over extended periods of time as if they were random changes of no selective importance to the species in which they occur.

Grassé has argued against Darwinian evolution. Having said that only paleontologists can speak authoritatively about evolution, he turns to the most modish discussion of DNA, codons, mutation, and gene origins to support his own views of organic evolution. Unfortunately, Cuvier—an ardent antievolutionist—was a paleontologist; being a practitioner of that branch of biology does not guarantee one's belief in the evolutionary process. Equally unfortunate, Grassé's grasp of the relationship between phenotype (especially in the sense of morphological structures) and genotype is woefully inadequate. If someone were to find a mutant allele which caused an imperfection in the iris of the eye, Grassé would seemingly conclude that the pupil of the eye is caused by a gene. He fails so badly in separating the notions of "what is necessary" and "what is sufficient" in respect to gene action that the thesis he presents in his recent book (Grassé 1973, 1977) can be dismissed.

Critics of Darwinian evolution, especially those interested in the origins of man, are prone to point out the gaps in our knowledge. Of course there are gaps in our knowledge! The science of genetics dates from the rediscovery in 1900 of Mendel's 1865 paper on inheritance in the garden pea; genetics is less than a century old. Protoplasmic substances which generations of biologists considered to be slimy nonspecific chemicals (like "egg white") are now known to have compositions as precise as any of the simple substances studied by first-year chemistry students. How is such precise specificity conferred during the synthesis of these protein molecules? How is the information or the instructions for making these enormous molecules stored in each egg and sperm? The attack on these and similar problems became possible only within the past decade or two. How unwise it would be for any evolutionist to deny that gaps exist in the data bearing on evolution. How unwise it would be for any antievolutionist to assume that today's gaps will persist through tomorrow. And, again, how unwise for anyone—Darwinian or anti-Darwinian—to believe that he alone has gained sufficient wisdom to foretell the future of biological science in all its fine detail.

Because of the interplay of its biological and cultural evolutions, the human species has arrived at a critical time in its history. To illustrate how critical today's events are, one might revert to an analogy: Human beings, like all vertebrates, have an elaborate immune system that provides the individual's defense against invasion by bacteria, viruses, or other alien substances. Let a strange microbe be encountered within an individual's body by the proper white blood cell, and a virtual cascade of immunological events set in motion—events aimed at overwhelming and destroying this organism lest it prove to be a threat to life.

The cascade of events that is set in motion is so dangerous to the life of the individual, however, that at the same moment another cascade of events is set in motion, designed to *turn off* the first series of immunological reactions. The successful defense against a foreign organism involves an interplay between two systems—one providing the defense, the second neutralizing the first—both of which operate at tremendous speed. Should the invading microbe be reproducing, its reproduction provides a third variable. The lives of higher organisms, consequently, depend on the interactions of two, even three, contrasting systems, each proceeding at tremendous speed.

Man's cultural evolution, like an individual's production of antibodies, is now moving at a horrendous pace. By examining the feedback mechanisms that have made biological evolution possible, we can use the logical countermeasures, the feedback, that will limit and control some of the more dangerous aspects of our culture. Unfortunately, the speed with which countermeasures can be devised does not always correspond to that at which cultural disharmonies arise.

"Do not do irreversible harm" is a pious statement of cautionary intent often used in guiding new technologies and their proponents. Increasingly often, however, by the time the statement has been uttered the damage has been done and new problems, new technological procedures, and new proponents wait to be addressed. And, once again, the time it takes to give the warning—"Do not do. . . ."—is too long to make the warning effective.

We do not know whether mankind will successfully develop the feedback mechanisms needed to control and guide cultural evolution. Perhaps the exhaustion of the earth's once colossal stores of fossil fuels will slow events to manageable speeds. Perhaps the development of a new ethic to replace those based on family and tribe will bring about the necessary change. By whatever means the change is brought about (if, indeed, it is) two groups of persons will have provided precious little help: those who claim that our fate lies in God's hands—let Him do as He may see fit; and those who merely say, "We know not what to do."

Glossary

Allele: A particular form of a gene. For example, either the *normal* or the *mutant* allele of a gene that, as a rule, specifies a functional enzyme; the mutant allele may make a defective enzyme, or none at all.

Allopolyploidy: The possession of chromosomal complements having their origins in two (or more) different species.

Amphioxus: A simple vertebrate-like organism that lacks a vertebral column and other bones but possesses a cartilaginous rod, a notochord, that serves to stiffen its body.

Autogenesis: Determination of the course of evolution from within organisms according to a preset (usually hypothetical or postulated) program.

Balancing selection: Selection acting on the frequencies of alleles in a population in which heterozygous individuals possess the highest average fitness.

Cyclostome: A primitive, jawless fish that is eellike in appearance and preys on (or is parasitic on) true fish.

Cytochrome: One of several pigmented proteins that function in photosynthesis and cellular respiration.

Dizygotic twins: Twins arising from two fertilized eggs; except for being the same age, such twins are no more similar than any sibling pair.

Dominant allele: An allele which has the ability to determine an individual's phenotype despite the presence of another (recessive) allele.

Echinoderm: The starfish and its relatives, such as the sea urchin and the sand dollar.

Electrophoresis: The separation of molecules by virtue of their different rates of migration through a supporting matrix in a high voltage electrical field.

Epigenesis: The theory that some physiological, developmental, or (in the minds of some persons) evolutionary processes, once set in motion, proceed without constant or specific genetic intervention.

Equifinality: The production, as the result of a stimulus, of a mutant gene whose product is identical to that evoked by the original stimulus itself, as would be the case if the brown pigment that is responsible for tan skin, acting as a stimulus, mutated the gene(s) whose action would be to increase the production of brown pigment.

Eukaryote: An organism whose genetic material is contained within a nucleus that possesses a nuclear membrane.

Fibrinopeptide: One of two short polypeptides that must be removed enzymatically from fibrinogen before the latter can aggregate to form a blood clot. Fibrinogen, the nonclotting precursor protein, circulates freely in the bloodstream.

Gene frequency: The frequency with which an allele is found among all individuals of a population. To be precise, this commonly used term should be "allele frequency." It can vary from 0 percent (no individual carries the allele in question) to 100 percent (all individuals are homozygous for it).

Gene locus: The site on a chromosome at which one or another allele of a given gene is to be found.

Genetic load: In lay terms, the burden imposed on a population by the presence of defective, mutant alleles; in mathematical terms, the amount by which the average fitness (or other measurable aspect) of a population is lowered relative to that of individuals possessing the maximum or optimum genotype.

Genome: A single complete set of genes or chromosomes.

Genotype: The constitution of an individual with respect to genetic endowment, either at all gene loci or at specified loci.

Hemoglobin: The protein of red blood cells that transports oxygen from the lungs to various tissues. The hemoglobin of many higher animals (including man) is an aggregate of four polypeptide chains: two alpha chains and two beta chains.

Heterosis: The possession by a hybrid individual of a phenotype that is maximal in some sense relative to those of its nonhybrid counterparts. Higher yield in crop plants and stamina in the mule are examples.

Heterozygous: Possessing two different alleles at a given gene locus.

Hominal primate: A primate recognized as belonging to the evolutionary branch leading to human beings.

Homologous: Similar by virtue of descent; used in contrast to *analogous,* which means similar by virtue of function or appearance. Wings of birds and bats are homologous; wings of birds and moths are analogous.

Homozygous: Possessing two identical alleles at a given gene locus.

Hybrid: An individual whose parents were members of different populations, varieties, races, species, or even genera. The degree of hybridity is often indicated, as in the term "interracial hybrid."

Infraspecific: Within a species, or at a taxonomic level below that of the species.

Lancelet: see *Amphioxus.*

Locus: see *Gene locus.*

Monozygotic twins: Twins arising from a single fertilized egg. Such twins are identical genetically and are often called "identical twins."

Neural tube: A hollow tube, formed by a longitudinal infolding of the developing embryo's outer layer of cells (ectoderm), that gives rise to the central nervous system.

Nomogenetic: Development (individual or evolutionary) that is said to proceed according to existing (usually unknown or postulated) laws.

Notochord: The evolutionary precursor of the vertebral column; a cartilaginous rod that stiffens the bodies of certain lower animals such as *Amphioxus.*

Nucleotide: A chemical unit, consisting of a purine or pyrimidine, a sugar, and a phosphate, that (depending on the sugar) is an integral part of DNA or RNA.

Ontogeny: The developmental history of an individual, extending from conception to death.

Orthogenesis: The apparent course of evolutionary change along a preset path (as from small primitive organisms to large descendant ones), assumed by some to reveal an "inner force" directing organic evolution.

Phenotype: The appearance and other physical attributes of an individual, as opposed to that individual's genetic endowment. Many genotypes can lead to the same phenotype even within a constant environment; many phenotypes can result from a single genotype if its carriers develop in different environments.

Phylogeny: The evolutionary history of an organism or group of organisms.

Pleiotropy: Modification of many aspects of an individual's phenotype resulting from the substitution of one allele for another. The mutant allele that "causes" white eyes in *Drosophila*, for example, alters as well the pigmentation of larval mouth parts and of the adult testes and associated structures, lowers viability, and interferes with mating behavior.

Polygene: A gene whose effect on some aspect of the individual's phenotype is small compared to that of the environment; thus, the phenotype for such a gene is an extremely poor indicator of the individual's genotype.

Polymorphism: The existence of two or more forms (morphs) in a single population.

Polypeptide chains: A linear sequence of amino acids which either by itself or in combination with other polypeptide chains constitutes a protein. Short polypeptide chains of a dozen or more amino acids often serve as hormones or other molecular signals.

Polyploidy: The possession of a set of chromosomes resulting from the doubling (tetraploidy), tripling (hexaploidy), or higher multiplication of the standard (diploid) set.

Pongial primate: A primate recognized as belonging to an evolutionary branch leading to one of the great apes.

Prokaryote: A primitive organism, usually a single cell, that lacks a nuclear membrane.

Recessive allele: An allele which in the presence of a dominant allele has no apparent effect on an individual's phenotype, which will reflect the action of the dominant allele.

Scutellum: A small triangular portion of a fly's thorax; it normally carries four large (scutellar) bristles.

Stochastic: Involving discrete events such that one can speak only of the probability that one or the other event will occur. For example, one half of all newborn babies are male, but for any given birth the child is either male *or* female, although the probability of its being either sex was 50 percent.

Synapse: The junction of two nerve cells or of a nerve cell and its target cell (for example, a muscle cell).

Tunicate: A simple organism that lacks a vertebral column (and all other bones as well) and possesses only a partial notochord (see also *Amphioxus*).

References

Ayala, F. J. 1968. Biology as an autonomous science. *Amer. Sci.* 56:207–221.

Ayala, F. J. 1974. The concept of biological progress. In F. J. Ayala and Th. Dobzhansky, eds., *Studies in the Philosophy of Biology: Reductionism and Related Problems*. Berkeley: University of California Press.

Ayala, F. J., J. R. Powell, M. L. Tracy, C. A. Mourão, and S. Pérez-Salas. 1972. Enzyme Variability in the *Drosophila willistoni* Group. IV. Genic variation in natural populations of *Drosophila willistoni*. *Genetics* 70:113–139.

Ayala, F. J., M. L. Tracy, L. G. Barr, J. F. McDonald, and S. Pérez-Salas. 1974. Genetic variation in natural populations of five *Drosophila* species and the hypothesis of selective neutrality of protein polymorphisms. *Genetics* 77:343–384.

Barghoorn, E. S. and J. W. Schopf. 1966. Microorganisms three billion years old from the Precambrian of South Africa. *Science* 152:758–763.

Berg, L. S. 1969. *Nomogenesis, or Evolution Determined by Law*. Cambridge: M.I.T. Press.

Bernardin de Saint-Pierre, J. H. 1784. *Etudes de la Nature*. Paris: P. F. Didot.

Bernstein, F. 1925. Zusammenfassende Betrachtungen über die erblichen Blutstrukturen des Menschen. *Zeitschr. Abstgs.- u. Vererbgsl.* 37:237–270.

Birch, L. C. 1974. Chance, necessity, and purpose. In F. J. Ayala and Th. Dobzhansky, eds., *Studies in the Philosophy of Biology: Reductionism and Related Problems* Berkeley: University of California Press.

Bose, N. K. 1951. Caste in India. *Man in India* 31:107–123.

Boule, M. 1946. *Les Hommes Fossiles: Eléments de Paléontologie Humaine*. Paris: Masson. Eng. trans. Boule, M. and H. V. Vallois. *Fossil Men*. New York: Dryden Press. 1957.

Bournoure, L. 1939. *L'Origine des Cellules Reproductrices et le Problème de la Lignèe Germinale*. Paris: Gauthier-Villars.

Brehm, A. E. 1866. *Illustrirtes Thierleben*. Vol. 3. Hildburghausen: Bibliogr. Institut.

Büchner, L. 1869. *Conférences sur la Théorie Darwinienne de la Transmutation des Espèces et de l'Apparition du Monde Organique*. Leipzig: T. Thomas.

Buffon, G. L. de. 1749. *Histoire Naturelle.* Vol. 1. Paris: Impr. Royale.

Bullini, L., V. Sbordoni, and P. Ragazzini. 1969. Mimetismo mülleriano in popolazione italiane di *Zygaena ephialtes* (L.) (Lepidoptera, Zygaenidae). *Arch. Zool. Ital.* 44:181–214.

Campbell, D. T. 1972. On the genetics of altruism and the counterhedonic components of human culture. *Jour. Social Issues* 28:21–37.

Cavalli-Sforza, L. L. and W. F. Bodmer. 1971. *The Genetics of Human Populations.* San Francisco: Freeman.

Clegg, M. T. and R. W. Allard. 1972. Patterns of genetic differentiation in the slender wild oat species *Avena barbata. P.N.A.S., U.S.* 69:1820–1824.

Critchley, M. 1960. The evolution of man's capacity for language. In S. Tax, ed., *Evolution after Darwin: The Evolution of Man.* Chicago: University of Chicago Press.

Cuenot, L. 1941. *Invention et Finalité en Biologie.* Paris: Flammarion.

Cuvier, G., Baron. 1825. *Discours sur les révolutions de la surface du globe, et sur les changements qu'elles ont produits dans la règne animal.* Paris: Dufour et d'Ocagne.

Cuvier, G., Baron. 1978. *Essay on the Theory of the Earth.* New York: Arno Press (original: 1817).

Darwin, C. 1859. *On the Origin of Species by Means of Natural Selection.* London: John Murray.

Darwin, C. 1871. *The Descent of Man and Selection in Relation to Sex.* London: John Murray.

Darwin, C. and A. R. Wallace. 1858. On the tendency of species to form varieties; and on the perpetuation of varieties and species by natural means of selection. *Jour. Linnean Soc. London, Zoology* 3:45–62.

Delsol, M. 1973. *Hasard, Ordre, et Finalité.* Quebec: Université Laval.

Descoqs, P. 1944. *Autour de la Crise du Transformisme.* Paris: Beauchesne et ses fils.

Diderot, D. and J. d'Alembert. 1751. *Encyclopédie, ou, Dictionnaire Raisonné des Sciences, des Arts et des Métiers.* Paris: Braisson.

Dissanayake, E. 1974. A hypothesis of the evolution of art from play. *Leonardo,* 7:1–7.

Dobzhansky, Th. 1943. Genetics of Natural Populations. IX. Temporal changes in the composition of populations of *Drosophila pseudoobscura. Genetics* 28:162–186.

Dobzhansky, Th. 1954. Evolution as a creative process. *Proc. 9th Int. Cong. Gen.* 1:435–449.

Dobzhansky, Th. 1970. *Genetics of the Evolutionary Process.* New York: Columbia University Press.

Dobzhansky, Th. 1973a. Ethics and values in biological and cultural evolution. *Zygon* 8:261–281.

Dobzhansky, Th. 1973b. *Genetic Diversity and Human Equality.* New York: Basic Books.

Dobzhansky, Th. and F. J. Ayala. 1973. Temporal frequency changes of enzyme and chromosomal polymorphisms in natural populations of *Drosophila*. *P.N.A.S.*, *U.S.* 70:680–683.

Dobzhansky, Th. and E. Boesiger. 1968. *Essais sur l'Evolution*. Paris: Masson.

Dobzhansky, Th., O. Pavlovsky, and J. R. Powell. 1976. Partially successful attempt to enhance reproductive isolation between semispecies of *Drosophila paulistorum*. *Evolution* 30:201–212.

Du Bois-Reymond, E. 1872. *Über die Grenzen des Naturekennens*. Leipzig.

Eckland, B. K. 1967. Genetics and sociology, a reconsideration. *Amer. Sociol. Review* 32:173–194.

Epstein, R., R. P. Lanza, and B. F. Skinner. 1981. "Self-awareness" in the pigeon. *Science* 212:695–696.

Erlenmeyer-Kimling, L. and L. F. Jarvik. 1963. Genetics and intelligence: a review. *Science* 142:1477–1479.

Eysenck, H. J. 1971. *Race, Intelligence, and Education*. New York: Library Press.

Ford, E. B. 1971. *Ecological Genetics*. London: Chapman and Hall.

Frisch, K. von. 1914. Der Farbensinn und Formensinn der Biene. *Zool. Jahrb. Physiol.* 37:1–238.

Gagnebin, E. 1943. *Le Transformisme et l'Origine de l'Homme*. Lausanne: F. Rouge.

Gardner, R. A. and B. T. Gardner. 1969. Teaching sign language to a chimpanzee. *Science* 165:664–672.

Goldschmidt, R. 1940. *The Material Basis of Evolution*. New Haven: Yale University Press.

Gould, J. 1865. *Handbook to the Birds of Australia*. London: J. Gould.

Grant, V. 1963. *The Origin of Adaptations*. New York: Columbia University Press.

Grassé, P. P. 1973. *L'Evolution du Vivant*. Paris: Albin Michel.

Grassé, P. P. 1977. *Evolution of Living Organisms*. New York: Academic Press.

Haeckel, E. 1882. *Die Naturanschauung von Darwin, Goethe, und Lamarck*. Jena: C. Fischer.

Haldane, J. B. S. 1932. *The Causes of Evolution*. New York: Harper and Row. Rpt. 1966. Ithaca, N.Y.: Cornell University Press.

Hamilton, W. D. 1964. The genetical evolution of social behavior. I and II. *Jour. Theoretical Biol.* 7:1–52.

Hamrick, J. L. and R. W. Allard. 1972. Microgeographical variation in allozyme frequencies in *Avena barbata*. *P.N.A.S.*, *U.S.* 69:2100–2104.

Hardin, G. 1963. A second sermon on the mount. *Persp. in Biol. and Med.* 6:366–371.

Hardin, G. 1970. Can teachers tell the truth about population? In A. B. Grobman, ed., *Social Implications of Biological Education* Washington: Natl. Assn. Biol. Teachers.

Hayes, K. J. and C. Hayes. 1954. The cultural capacity of chimpanzee. *Human Biol.* 26:288–303.

Hearnshaw, L. S. 1979. *Cyril Burt, Psychologist*. Ithaca, N.Y.: Cornell University Press.

Huxley, J. S. 1941. *Man Stands Alone.* New York: Harper and Bros.

Huxley, T. H. 1863. *Evidence as to Man's Place in Nature.* London: William and Norgate.

Jacob, F. 1970. *La Logique du Vivant: Une Histoire de l'Hérédité.* Paris: Gallimard.

Jacob, F. 1976. *The Logic of Life: A History of Heredity.* New York: Random House, Vintage Books.

Jacob, F. 1977. Evolution and tinkering. *Science* 196:1161–1166.

Jencks, C. 1973. *Inequality: A Reassessment of the Effect of Family and Schooling in America.* New York: Harper and Row.

Jensen, A. R. 1969. How much can we boost IQ and scholastic achievement? *Harvard Ed. Rev.* 39:1–123.

Jensen, A. R. 1973. *Educability and Group Differences.* London: Methuen.

Johannsen, W. 1909. *Elemente der exakten Erblichkeitslehre.* Jena: Gustav Fischer.

Johnson, F. M. 1971. Isozyme polymorphism in *Drosophila ananassae:* Genetic diversity among island populations in the South Pacific. *Genetics* 68:77–95.

Kammerer, Paul. 1924. *The Inheritance of Acquired Characteristics.* New York: Boni and Liveright.

Kimura, M. 1968. Evolutionary rate at the molecular level. *Nature* 217:624–626.

Kimura, M. and J. F. Crow. 1964. The number of alleles that can be maintained in a finite population. *Genetics* 49:725–738.

Kimura, M. and T. Ohta. 1971. *Theoretical Aspects of Population Genetics.* Princeton, N.J.: Princeton University Press.

King, J. L. and T. H. Jukes. 1969. Non-Darwinian evolution: Random fixation of selectively neutral mutations. *Science* 164:788–798.

Koehn, R. K. and D. J. Rasmussen. 1967. Polymorphic and monomorphic serum esterase heterogeneity in catastomid fish populations. *Biochem. Genetics* 1:131–144.

Lamarck, J. B. 1802. *Recherches sur l'Organisation des Corps Vivants.* Paris.

Lamarck, J. B. 1809. *Philosophie Zoologique* (Edition Nouvelle, Paris: Chas. Martins, Librairie Savy, 1873).

Lamarck, J. B. 1914. *Zoological Philosophy: An Exposition with Regard to the Natural History of Animals.* London: Macmillan.

Leroi-Gourhan, A. 1965. Préhistoire de l'art occidental. Paris: Editions d'Art Lucien Maxenod.

Lerner, I. M. 1954. *Genetic Homeostasis.* Edinburgh: Oliver and Boyd.

Lewontin, R. C. 1974. *The Genetic Basis of Evolutionary Change.* New York: Columbia University Press.

Lewontin, R. C. and J. L. Hubby. 1966. A Molecular Approach to the Study of Genic Heterozygosity in Natural Populations. 2. Amount of variation and degree of heterozygosity in natural populations of *Drosophila pseudoobscura. Genetics* 54:595–609.

Lyell, Sir Charles. 1863. *The Geological Evidences of the Antiquity of Man.* London: J. Murray.

Malraux, A. 1960. *L'Univers des Formes.* Paris: Gallimard.

Malthus, T. R. 1806. *An Essay on the Principle of Population*. London: J. Johnson.

Marshall, A. J. 1954. *Bower Birds, Their Displays and Breeding Cycles*. Oxford: Clarendon Press.

McDonald, J. and F. J. Ayala. 1974. Genetic response to environmental heterogeneity. *Nature* 250:572–574.

Monod, J. 1970. *Le Hasard et la Nécessité*. Paris: Seuil.

Monod, J. 1971. *Chance and Necessity*. New York: Knopf.

Morgan, T. H. 1916. *A Critique of the Theory of Evolution*. Princeton, N.J.: Princeton University Press.

Morris, D. 1962. *The Biology of Art*. London: Methuen.

Mukai, T. 1964. The Genetic Structure of Natural Populations of *Drosophila melanogaster*. I. Spontaneous mutation rate of polygenes controlling viability. *Genetics* 50:1–19.

Muller, H. J. 1950. Our load of mutations. *Amer. Jour. Human Genetics* 2:111–176.

Nagel, E. 1961. *The Structure of Science*. New York: Harcourt, Brace and World.

Osborn, H. F. 1917. *The Origin and Evolution of Life*. New York: Scribner.

Osborn, H. F. 1921. *L'Origine et l'Evolution de la Vie*. Paris: Masson.

Paley, W. 1802. *Natural Theology*. London: R. Faulder.

Portmann, A. 1947. *Vom Bild der Natur*. Basel: F. Reinhardt.

Powell, J. R. 1971. Genetic polymorphisms in 'varied environments. *Science* 174:1035–1036.

Premack, D. 1971. Language in chimpanzee? *Science* 172:808–822.

Rensch, B. 1957. Ästhetische Faktoren bei Farb- und Formbevorzugungen von Affen. *Zeitschr. f. Tierpsychologie* 14:71–99.

Rensch, B. 1965. Uber ästhetische Faktoren im Erleben höherer Tiere. *Naturw. und Medizin*. 2:43–57.

Rensch, B. 1974. Polynomistic determination of biological processes. In F. J. Ayala and Th. Dobzhansky, eds., *Studies in the Philosophy of Biology: Reductionism and Related Problems* Berkeley: University of California Press.

Robert, P. 1951–1964. *Dictionnaire alphabétique et analogique de la langue française* (9 vols.). Paris: Presses Universitaires de France.

Romanes, G. J. 1892. *Darwin and after Darwin. I. The Darwinian Theory*. London: Longmans, Green and Co.

Salvini-Plawen, L. V. and E. Mayr. 1977. On the evolution of photoreceptors and eyes. *Evol. Biol.* 10:207–263.

Scarr-Salapatek, S. 1971a. Unknowns in the IQ equation. *Science* 174:1223–1228.

Scarr-Salapatek, S. 1971b. Race, social class, and IQ. *Science* 174:1285–1295.

Schopf, J. W. 1974. Paleobiology of the Precambrian: The age of the blue-green algae. *Evol. Biol.* 7:1–43.

Selander, R. K., M. H. Smith, S. Y. Yang, W. E. Johnson, and J. B. Gentry. 1971. Biochemical polymorphisms and systematics in the genus *Peromyscus*. I. Variation in the old-field mouse. *Univ. Texas Publ.* 7103:49–90.

Selander, R. K., S. Y. Yang, R. C. Lewontin, and W. E. Johnson. 1970. Genetic variation in the horseshoe crab (*Limulus polyphemus*), a phylogenetic "relic." *Evolution* 24:402–414.

Selous, E. 1929. Schaubalz und geschlechtliche Auslese beim Kampflaufer *Philomachus pugnax. J. F. Ornithol.*, 77:262.

Sherrington, C. 1953. *Man and his Nature.* Garden City, N.Y.: Doubleday Anchor.

Simpson, G. G. 1964. *This View of Life.* New York: Harcourt, Brace, and World.

Simpson, G. G. 1969. *Biology and Man.* New York: Harcourt, Brace, and World.

Spieth, H. T. 1968. Evolutionary implications of sexual behavior in *Drosophila. Evol. Biol.* 2:157–193.

Sturtevant, A. H. 1965. *A History of Genetics.* New York: Harper and Row.

Teilhard de Chardin, P. 1955. *Le Phénomène Humain.* Paris: Seuil.

Teilhard de Chardin, P. 1959. *The Phenomenon of Man.* New York: Harper.

Timofeeff-Ressovsky, N. W. 1940. Zur Analyse des Polymorphismus bei *Adalia bipunctata* L. *Biol. Zbl.* 60:130–137.

Valéry, P. 1968. *Oeuvres, Théorie Poétique et Esthétique.* Vol. 1. Paris: Bibliothèque La Pléiade.

Verheyen, R. 1950. *Les Colombidés et les Gallinacés.* Brussels: Inst. Royal des Sci. Nat. de Belgique.

Vogt, C. 1863. *Vorlesungen über den Menschen, seine Stellung in der Schöpfung und in der Geschichte der Erde.* Giessen: J. Rickersche.

Wallace, B. 1968. *Topics in Population Genetics.* New York: W. W. Norton.

Wallace, B. 1981. *Basic Population Genetics.* New York: Columbia University Press.

Wallace, B., M. W. Timm, and M. P. P. Strambi. 1983. The establishment of novel mate-recognition systems in introgressive hybrid *Drosophila* populations. *Evol. Biol.* 16 (in press).

Weismann, A. 1892. *Das Keimplasma.* Jena: G. Fischer.

Weismann, A. 1893. *The Germ Plasm.* New York: Charles Scribner's Sons.

Wilson, A. C. and V. M. Sarich. 1969. A molecular time scale for human evolution. *P.N.A.S., U.S.* 63:1088–1093.

Wright, S. 1931. Evolution in Mendelian populations. *Genetics* 16:97–159.

Wright, S. 1949. Adaptation and selection. In G. L. Jepsen, G. G. Simpson, and E. Mayr, eds., *Genetics, Paleontology, and Evolution.* Princeton, N.J.: Princeton University Press.

Index